AF684756

10 Jan. 14
Oxford University
Karachi

Institutional Barriers to Sustainable Urban Transport in Pakistan

Institutional Barriers to Sustainable Urban Transport in Pakistan

MUHAMMAD IMRAN

Foreword by
NICHOLAS LOW

OXFORD
UNIVERSITY PRESS

OXFORD
UNIVERSITY PRESS

Great Clarendon Street, Oxford OX2 6DP

Oxford University Press is a department of the University of Oxford.
It furthers the University's objective of excellence in research, scholarship,
and education by publishing worldwide in

Oxford New York

Auckland Cape Town Dar es Salaam Hong Kong Karachi
Kuala Lumpur Madrid Melbourne Mexico City Nairobi
New Delhi Shanghai Taipei Toronto

with offices in

Argentina Austria Brazil Chile Czech Republic France Greece
Guatemala Hungary Italy Japan Poland Portugal Singapore
South Korea Switzerland Turkey Ukraine Vietnam

Oxford is a registered trade mark of Oxford University Press
in the UK and in certain other countries

© Oxford University Press 2010

The moral rights of the author have been asserted

First published 2010

All rights reserved. No part of this publication may be reproduced, translated,
stored in a retrieval system, or transmitted, in any form or by any means,
without the prior permission in writing of Oxford University Press.
Enquiries concerning reproduction should be sent to
Oxford University Press at the address below.

This book is sold subject to the condition that it shall not, by way
of trade or otherwise, be lent, re-sold, hired out or otherwise circulated
without the publisher's prior consent in any form of binding or cover
other than that in which it is published and without a similar condition
including this condition being imposed on the subsequent purchaser.

ISBN 978-0-19-547666-8

Typeset in Adobe Garamond Pro
Printed in Pakistan by
Kagzi Printers, Karachi.
Published by
Ameena Saiyid, Oxford University Press
No. 38, Sector 15, Korangi Industrial Area, PO Box 8214
Karachi-74900, Pakistan.

To my mother and father who taught me to think differently
and
to Ayesha, Zainab and Amnah whose love put it all together

Contents

List of Figures and Tables

Figures

Tables

List of Abbreviations

ADB	Asian Development Bank
ADP	Annual Development Programme
ALGAS	Asian Least-cost Greenhouse Gas Abatement Strategy
APCC	Annual Plan Coordination Committee
ASPO	Association for the Study of Peak Oil
BOO	Build, Own and Operate
BOOT	Build, Own, Operate and Transfer
BOT	Build, Operate and Transfer
BP	British Petroleum
C&W	Communication and Works
CBD	Central Business District
CCB	Citizen Community Boards
CDG	City District Government
CDWP	Central Development Working Party
CENTO	Central Treaty Organisation
CG	City/District Government
CIDA	Canadian International Development Agency
CILT	Chartered Institute of Logistics and Transport
CITP	Chartered Institute of Transport of Pakistan
CNG	Compressed Natural Gas
CO	Carbon Monoxide
CO2	Carbon Dioxide
CSTS	Comprehensive Study on Transportation System
DDWP	Departmental Development Working Party
DETR	Department of Environment, Transport and Regions (UK)
DPL	Development Policy Loan
EA	Environmental Accounting
ECC	Economic Co-ordination Committee
ECNEC	Executive Committee of National Economic Council
EIA	Environmental Impact Assessment
EIM	Environmental Information Management
ENERCON	Pakistan National Energy Conservation Centre
Engg.	Engineering

EPA	Environmental Protection Authority
EPD	Environmental Protection Department
EUAD	Environment and Urban Affairs Division
FETRS	Fuel Efficiency in Road Transport Sector Project
FWO	Frontier Works Organisation
GDP	Gross Domestic Product
GHG	Greenhouse Gas
GNP	Gross National Product
GoP	Government of Pakistan
GPA	Gwadar Port Authority
GT Road	Grand Trunk Road
HI	Historical Institutionalism
HRD	Hardware
IBRD	International Bank for Reconstruction and Development
IDA	International Development Agency
IEE	Initial Environmental Examination
IMF	International Monetary Fund
IPCC	Intergovernmental Panel on Climate Change
IUCN	International Union for Conservation of Nature
JICA	Japan International Cooperation Agency
KCR	Karachi Circular Railway
KPT	Karachi Port Trust
LCC	London County Council
LCDG	Lahore City District Government
LDA	Lahore Development Authority
LHC	Lahore High Court
LMA	Lahore Metropolitan Area
LPG	Liquid Petroleum Gas
LRRP	Lahore Roads Rehabilitation Project
LRT	Light Rail Transit
LRTA	Lahore Regional Transport Authority
LUDTS	Lahore Urban Development and Traffic Study
LUTP	Lahore Urban Transport Project
MCL	Metropolitan Corporation Lahore
MMBW	Melbourne Metropolitan Board of Works
MMP	Mott MacDonald Pakistan
MTDF	Medium Term Development Framework
MTR	Mid-term Review
MVE	Motor Vehicle Examiners
NCS	National Conservation Strategy

NEAP	National Environmental Action Plan
NEC	National Economic Council
NEQs	National Environmental Quality Standards
NESPAK	National Engineering Services Pakistan
NGO	Non Government Organisation
NHA	National Highway Authority
NHB	National Highway Board
NHC	National Highway Council
NIT	National Institute of Transportation
NO2	Nitrogen Oxide
NTB	National Transport Board
NTRC	National Transport Research Centre
NUST	National University of Science and Technology
OCAC	Oil Companies Advisory Committee
ODA	Overseas Development Assistance
P&D	Planning and Development
PACO	Pakistan Automobile Corporation
Pak-EPA	Pakistan Environmental Protection Authority
PC1	Project Concept 1
PCU	Passenger Car Unit
PDWP	Provincial Development Working Party
PEC	Pakistan Engineering Council
PEPA	Pakistan Environmental Protection Act
PEPO	Pakistan Environmental Protection Ordinance
PFC	Provincial Finance Commission
PHA	Parks and Horticulture Authority
PHATA	Punjab Housing and Town Planning Agency
PIDE	Pakistan Institute of Development Economics
PNSC	Pakistan National Shipping Corporation
PQA	Port Qasim Authority
PRTC	Punjab Road Transport Corporation
PSDP	Public Sector Development Programme
PTA	Provincial Transport Authority
PUDP	Punjab Urban Development Project
PUTP	Provincial Urban Transport Policy
PWD	Public Works Department
R&B	Roads and Bridges
RCI	Rational Choice Institutionalism
ROW	Right-of-Ways
RPAA	Regional Planning Association of America

RTA	Regional Transport Authority
SACTRA	Standing Advisory Committee on Trunk Road Assessment
SCI	Social Constructive Intuitionalism
SDO	Sub-divisional Officer
SDPI	Sustainable Development Policy Institute
SE	Superintending Engineer
SEA	Strategic Environmental Assessment
SEATO	South East Asian Treaty Organisation
SIA	Social Impact Assessment
SIDA	Swedish International Development Agency
SME	Small and Medium Enterprises
SMEDA	Small and Medium Enterprise Development Authority
SO2	Sulphur Dioxide
SPM	Suspended Particulate Matter
TC	Transport Council
TCC	Transport Coordination and Cooperation (Committee)
TEPA	Traffic Engineering and Transport Planning Agency
TEPU	Traffic Engineering and Transport Planning Unit
TMA	Tehsil Municipal Administration
TPB	Transport Planning Bureau
TSDI	Transport Sector Development Initiative
UET	University of Engineering and Technology
ug/m3	Microgram per Cubic Metre (of air)
UNDP	United Nations Development Programme
UNFCC	United Nations Framework Convention for Climate Change
UPRC	Urban Policy Research Centre
USAID	United States Agency for International Development
USGS	United States Geological Services
UTPP	Urban Transport Planning Process
V/C Ratio	Volume to Capacity Ratio
WHO	World Health Organisation
X.E.N	Executive Engineer
ZM	Zila Mohtasib

Foreword

It is no longer true that, as the French say, 'the more things change, the more they remain the same'. In the next hundred years nothing is going to remain the same. Fundamental change in our natural environment, which is the very foundation of our economies and societies, is already forcing itself upon us, and will grow with devastating impact.

In developing countries we have seen air pollution from motor vehicles rise unchecked, damaging public health in cities. Oil supplies are reaching a peak, while, until the global financial crisis hit, the growing demand for oil pushed prices to record levels. Oil prices have receded as demand has fallen, but will rise again to new heights once growth resumes. Climate change is now inevitable. The poor of the developing world will be the first to pay, and will pay the most. The greenhouse gas emitted from transport systems worldwide is a major and growing contributor to climate change. So the Western-inspired model of total motorized mobility is no longer viable even in the West. It cannot and must not continue to be transmitted to the developing world.

This book by Muhammad Imran sounds a call for the truth, to acknowledge that a country like Pakistan can never attain the dream of universal, individual motorized mobility projected so tantalizingly from the USA and other developed nations, and the aid agencies these countries support (directly or indirectly). Imran's vital message is that it is time for Pakistan to find its own way to mobility, to acknowledge the reality of how the masses move around their cities, and to cater for today's mass needs for collective and active travel, and not how they might be in some impossible dream of the future. The message is also for the international aid agencies. They must acknowledge that where Pakistan needs help is in developing the capacity to plan for its own transport future, and to rely less on their prescribed solutions.

Imran carefully lays out a detailed institutional history of transport policy-making in Pakistan and shows how the paradigm of large scale road building and associated infrastructure to accommodate private car traffic has become dominant in the large cities. That paradigm, imported from the developed world, and now itself signifying 'development' as defined by Western experts, has supplanted the development of integrated rail and bus systems, and planning for safe walking and cycling. How is it that when more than 50 per cent of all journeys in Lahore are made on foot, it is still not possible to walk in comfort and safety? The main concern, he says, has been to increase the amount of travel by car, while ignoring the quality of travel on public transport and non-motorized transport.

Imran's theory is that of institutional path dependence. Path dependence is a form of irrationality in which futures are decided by conventional wisdoms, rigid institutional structures and discursive rationalizations rather than by attempts to address real problems and meet real needs. Path dependence is determined by ways of thought that fail to respond to emergent problems like climate change, air pollution and peak oil. In 2008, path dependence brought the global financial world to the brink of ruin. Climate change is a slower moving event, but one which will be infinitely more dangerous than the financial crisis.

In the short term, it is the price of oil that makes planning a motorized future for Pakistan particularly irrational. Oil production has been predicted to peak within five-years. Once demand for oil from the developed world (particularly the USA) resumes its rise, its price will again be on an endless upward spiral. Poorer countries are especially at risk. Policies that increase car dependence will leave Pakistan vulnerable to reduced mobility.

In the long term, unless the world changes its model of motorized mobility, Pakistan will suffer from climate change in a particular and horrifying way. The Indus basin feeds Pakistan and is sustained during the hot summer months by flows from 3500 glaciers in the Himalayas. Average global warming of just three degrees will eventually melt the glaciers. The Indus valley has an arid climate and is dependent on glacial melt. Mark Lynas (2008) has this to say: 'The combination of aridity and a highly glaciated catchment makes the Indus particularly dependent on mountain water. And this is its Achilles heel: the once timeless mountain snows are already melting. A quirk of atmospheric dynamics means that the air temperatures in the highland region are rising at twice the global average, and accelerated glacial retreat has been recorded across all the major ice-capped ranges'. The Indus will first swell with the additional melt water. But once the ice has gone, the Indus will dry up. Unless global warming is contained below two degrees, Pakistan's future will be limited to not much more than two hundred years.

Pakistanis are not to blame for peak oil or climate change. Their impact per person is many times smaller than that of the developed economies. They have not consumed much of the world's oil resources. They have not poured forth greenhouse gas from fossil fuelled engines and industry. But they can no longer follow a fossil fuelled path to development. That path is illusory and leads not to the satisfaction of wealth but to famine and disaster for Pakistan.

This is a book not only for the developing world but also for the developed world as it seeks to assist the economic development of poorer countries. It challenges the conventional wisdom. It foreshadows a new and more truthful path ahead.

Professor Nicholas Low
Australasian Centre for the Governance and Management of Urban Transport (GAMUT)
The University of Melbourne, Australia

Acknowledgements

The basis of this book lies in my doctoral thesis completed at the University of Melbourne, Australia in 2006. The following are just a few of the people I would like to thank. My special thanks go to Professor Nicholas Low for his steady support and encouragement throughout my time at the University of Melbourne. I extend my gratitude to Prof. Kevin O' Conner who encouraged me to submit the draft manuscript to a publisher; and to Dr Paul Mees who is an 'activist academic' and always a source of inspiration for me.

I am grateful to all the academics, officials and other people in Pakistan who generously supported my study and subsequently provided me with necessary information. I would also like to acknowledge my doctoral thesis examiners and anonymous referees who provided supportive and useful comments.

I owe particular thanks to my colleagues at Massey University: Dr Henry Barnard, Dr Johanna Rosier, and Dr Marco Amati who recruited me and provided support to complete this book. I would also like to thank Oxford University Press who made the publication of this book possible.

Finally, this book could not have been written without the prayers of my parents, brothers and sisters, the support and encouragement of Ayesha, and the love of Zainab and Amnah.

Acknowledgements

Introduction

I grew up in the small city of Wah Cantt located approximately 50 kilometres north of Islamabad, Pakistan. From the beginning, I saw that almost everyone cycled to work. Similarly, every student went to school and college on foot. One would observe hundreds of cyclists and pedestrians in the early morning and afternoon on the Mall Road of Wah Cantt. Later, these journeys were complemented by a decent, reliable and efficient bus service for the residents of Wah Cantt. This was my first introduction to active (walking and cycling) and collective (public transport) modes of transport which mainly served work and educational purposes in urban areas.

In terms of inter-urban transport, I cannot recollect when I first started travelling to Lahore from Wah Cantt. However, I do remember that each year we spent the spring holidays in Lahore. We would travel by railcar (a railway vehicle consisting of a single, powered coach) which left Rawalpindi station early in the morning. Overall, travelling by railcar was pleasant, very comfortable and more or less predictable. I have no hesitation in stating that these journeys were the best intercity travelling experiences I have had in Pakistan.

From the Lahore railway station (a magnificent building with many open spaces) we preferred riding a *tonga* (a horse-driven cart which, in fact, was banned in Wah Cantt) to my grandfather's house. During our stay, we would ride buses and rickshaws (the three-wheeler rickshaws were also banned in Wah Cantt) but like other *Lahorees* (residents of Lahore), we preferred to hire a *tonga*, or use a cycle or simply go on foot. Our grandfather, who worked in the Pakistan Railway workshops in Mughalpura, would take us around to show us the vibrant city of Lahore which was well connected with active and collective modes of transport. The streets of Lahore were full of people, activities and fun. The air was clean and one could clearly see a blue sky. Thousands of people would go to their jobs on or around the Mall Road; hundreds more visited the shopping district of the Walled City and Anarkali; a large number of students attended schools and colleges situated all over Lahore and hundreds went to night cinemas on active and collective modes of transport. Even Model Town, an elite suburban neighbourhood, was connected to the main city by a bus service.

These memories developed my perceptions that a lively, clean and green city is possible when mainly served by active and collective modes of transport. These memories were later confirmed in Melbourne, Australia when I met an 85-year-old Indian lady—Mrs Nirmal Thakur, who had lived and studied in Lahore before the partition of the Indian subcontinent in 1947. She remembered Lahore's beauty during the 1930s and 1940s and told me how she would go to school and later college on foot. She shared her pleasant experience of using the Model Town bus service and of her walks along the Mall Road and the Lahore Canal. These modes of travelling made transport enjoyable and ultimately made Lahore more liveable. Such happy memories and experiences were the assets of Lahore which made the city unforgettable, even for those who migrated from it some 60 years ago, as in the case of Nirmal Thakur.

During my childhood, travelling within Lahore was a straightforward activity both for its citizens as well as visitors. Like other cities in the world, the fundamental and historical reason for the creation of Lahore was '*to reduce the need for travel*'. This was done by utilizing land in an organized manner so as to create separate residential, commercial, industrial and recreational areas. These areas were laid out such that each would benefit the other in generating activities and thus offer a higher quality of urban life. The fact that these historical considerations in the development of cities are now ignored result in the creation of added travel demand in dense and expensive urban areas. This demand can be dealt with by adopting the supply-side approach (widening and building more roads) or the demand-side approach (transport management in favour of walking, cycling and public transport). The problem of increased travel worsens when policy makers feel obliged to adopt the supply-side approach which is inevitably done at the cost of other land uses. Although this might appear to provide a quick solution to the problem of transport, it has the effect of making city life difficult and unpleasant, ultimately stripping cities of the qualities that made them liveable in the first place. This is particularly the case in developing countries which are facing massive pressure from population and traffic growth in densely populated cities.

The situation of urban transport has changed in Lahore over time. The *tonga* has been banned on environmental grounds and has been replaced by the Chingchee (motorcycle rickshaw). Omnibuses were replaced with Volvo buses, and then with ugly wagons and later franchised buses, all introduced without any investment and planning. It is interesting that local governments in cities of the developed world plan to increase population and employment density, whilst finding ways to decrease car ownership, encouraging non-motorized transport and reducing the need for travel for the development of sustainable urban transport systems. In the past Lahore too enjoyed all these characteristics and still does to a certain extent. However, the city's focus has now shifted to building wider and higher quality roads which include underpasses, flyovers and ring roads; this in turn has led to a decrease in population and employment density in the city, increased car ownership and commuting distances, and decreased possibility for non-motorized transport.

Most decision-makers, military or civilian, and the professionals responsible for urban transport own a car and hence their perception of the transport problem is largely from behind the 'windscreen' of a private vehicle. This perception is false as it does not represent the experience of the majority. These decision-makers present investments in road projects as being vital for solving urban transport issues but which unfortunately consume the bulk of the available transport funding and marginalize active and collective transport modes as well as their associated infrastructure. The results of these policies can be seen in the form of greater levels of air and noise pollution, higher petrol consumption and an increasing number of road accidents, ultimately making Lahore a difficult city to live in.

How are current policies exacerbating Lahore's transport problem and making the city unliveable? This book attempts to analyse the problem and provide some answers.

Time has clearly proven that the approach adopted for planning Lahore's transport has been unable to solve the problems which have increased over the years. A glimmer of hope emerged after the United Nations Earth Summit held in Rio de Janeiro in June 1992, which initiated the debate about sustainable development and declared that the ultimate aim was to have 'sustainable transport'. In general, sustainable transport means the system of transport which has a lower impact on the environment, society and economy. Transport systems in which the share of active and collective modes of transport is higher and continues to rise compared to private modes of transport (such as car and motorcycle) is internationally recognized as more sustainable. Pakistani cities, especially Lahore, have a higher share of active and collective modes of transport due to historical reasons but this share is decreasing very sharply. The Pakistani government signed the Rio declaration, and I believe that

Pakistani cities have the potential to fulfil the requirements of having sustainable transport. However, sixteen years after the Rio Conference, evidence has emerged that there were many conceptual lacunae and contradictory objectives in the planning and development of sustainable transport, especially in the case of Pakistan where transport development projects aim to enhance the sustainability agenda. But the important question is what does sustainable development really mean in the context of these projects? The author seeks to explore this pertinent question.

While writing this book, I found a number of best practice case studies in sustainable transport from around the world. One may produce a compendium of best practice models, but the crucial question is whether such practices would be successful in Pakistan? Often best practice models in sustainable development have failed due to barriers put up by institutions. Generally, formal or informal organizations and practices in the political or social life of a people are considered institutions. This book aims to investigate institutional barriers particularly in the context of urban transport in Pakistan. I selected urban transport because of its potential impact on greenhouse gas (GHG) emissions, energy use, the urban environment and road crashes. Thus, the overall aim of the book is to identify and examine institutional barriers that create a hindrance to sustainable development, and thereby prevent the application and implementation of sustainable urban transport policies in Pakistan. A close examination of institutions will make it clear whether or not they have advanced the commitment to sustainable development in the urban transport sector.

To analyse the process, this book will draw insights from the concept of path dependence which does not simply analyse the outcomes of urban transport solutions but examines all the factors that influence the way in which decision-makers conceive and address the problem of transport and mobility. I argue that developments in institutional politics critically depend on the specific social construction of the problem in its historical context. The study does not merely analyse what is being planned and implemented, but also critically evaluates the institutional context in which projects are proposed and perceived.

Policy making in urban transport is not simply a matter of finding acceptable solutions for pre-conceived problems, it is also the dominant way in which societies address and regulate new emerging agendas like sustainable development. However, policy making involves much more than creating solutions according to the different techniques and methods currently available. It requires re-conceiving the problem of urban transport in order to find solutions. The perception of problems and generating solutions become possible through institutional structures and arrangements. It would seem that the financial and technical influence of dominant institutions plays an important role in defining the problem and generating solutions.

This book investigates whether urban transport policy making is path dependent as a set of practices developed in history which now incorporate the sustainability agenda. In so doing, the following chapters seek to shed new light on the institutional context of policy making. My aim is to present a new perspective by investigating the relationship between international institutions, sources of finance and local problems. The goal is to highlight how these relationships should work, and how they are actually working in reality.

Throughout this book, the sustainability of urban transport would appear as a major challenge to the institutions of modern society. In light of this challenge, I will analyse policies and institutions dealing with the problem of urban transport in Lahore. This book raises questions about the international as well as national and local institutions' performance on the sustainability of urban transport. In term of outcomes, this book will explore the location of power and influence and how that power and influence has been exercised to promote or hinder the goal of sustainability in the urban transport sector. Finally, I will identify the kinds of barriers that will have to be overcome and the arguments that need to be reassessed in order for the new policy paradigm of sustainable urban transport to take root in Pakistan.

My research and professional experience in urban transport planning in Pakistani cities enabled me to write this book. I hope it provides an in-depth analysis and can be used as a reference by academics, and both local and international policy makers, for restoring the sustainability of urban transport in Pakistani cities. I accept any errors as mine and welcome feedback and opportunities to work collaboratively with other academics as well as organizations to make Pakistani cities and their transport systems more sustainable.

1

The Challenge of Sustainability: Urban Transport and Public Policy

Most people take the simple view that urban transport means movement of people and goods from one part of the city to another. This view overlooks an important aspect of transport which is that it not only serves the city but also shapes its environment and ultimately the quality of urban life. Traditionally, cities situated in territory that is now Pakistan grew on the basis of compact and diverse land uses accessible by foot. During the British rule, these cities gradually spread outward because of the introduction of modern means of transport such as trains, trams and buses. During this period, medium density housing and employment grew in a mixed land use along public transport corridors. More recently, however, low density housing has started to appear on the periphery of cities which has enhanced the use of cars. Since Pakistani cities retain their commercial and employment centres in the area traditionally served by walking, cycling and public transport, people who live in low density housing areas, beyond 20–30 km from the city centre, now have to use their cars to go for work. This pattern of travel generates enormous traffic at most times of the day and causes severe environmental problems in Pakistani cities. Therefore, the first issue that needs to be addressed is to determine the impact of urban transport on environment. The quality of life in Pakistani cities has declined over time, which seems to confirm the role of urban transport in the growing air and noise pollution, road crashes, the transport sector's energy consumption and greenhouse gas emissions. These growth pressures demand strong institutional efforts to reclaim our cities by delivering environment friendly solutions. However, institutions, especially in developing countries, appear to be creating hurdles in developing sustainable urban transport.

SUSTAINABILITY IN THE CONTEXT OF URBAN TRANSPORT

Over the past decade, the term 'sustainable development' has emerged as the principal concept in urban development in newly industrialized countries. Perhaps the most commonly accepted and popular definition of sustainable development is the one proposed by the Brundtland Report[1] which defined sustainable development as: 'Development which meets the needs of the present, without compromising the ability of future generations to meet their own needs.'

The Brundtland Commission adopted a multilateral and interdependent approach to environmental problems.[2] This concept involves not only the management and conservation of the natural resource base, but also the social, institutional, technological and cultural changes involved. Therefore, it is possible to say that sustainable development is the universally agreed goal of human progress. Such a goal requires a collective effort in the form of political will, technical ingenuity and a true 'paradigm shift' in the belief systems of the professionals who shape the city.[3] Since collective efforts are made possible through institutions, without first examining institutions we cannot move purposefully towards sustainability. The term 'institution' can be briefly defined as 'an underlying durable pattern of rules and behaviour'.[4] Although there has been much policy development under the banner of sustainable development in recent years, the institutional aspects remain at the margins of public policy analysis.[5]

Even though there is a coalition of interests supporting environmental sustainability, sustainable development is capable of multiple interpretations.[6] Lele has observed:

> Sustainable development is a 'metafix' that will unite everybody from the profit minded industrialist and risk-minimizing worker, the pollution concerned or wildlife loving First Worlder, the growth-maximizing policy maker, the goal-oriented bureaucrat, and therefore, the vote-counting politicians.[7]

Furthermore, Lele found that all the groups he mentions are using the same language and endorsing the same nominal objective that is 'sustainable development'.[8] However, a problem arises when different actors argue about how to achieve sustainable development. In this respect there is no clear theoretical and analytical framework. Jacobs argues that one consequence of this ambiguity is that not only does sustainable development permit multiple interpretations, but it also fails to provide a clear boundary between itself and the conventional understanding of successful development that it is intended to replace.[9] Therefore, in Jacobs view, sustainable development in practice is a 'contested' concept.[10] Luke has described the confusion about sustainable development, as:

> ...a concept fraught with unsolved questions: Sustainable for how long—a generation, one century, a millennium, ten millennia? Sustainable at what level of human appropriation: individual households, local villages, major cities, entire nations, and global economies? Sustainable for whom: all humans alive now, all humans that will ever live, all living beings living at this time, all living beings that will ever live? Sustainable under what conditions: for contemporary transnational capitalism, for low impact Neolithic hunters and gatherers, for some space-faring global empire? Sustainable development of what: personal income, social complexity, gross national product, material frugality, individual consumption, ecology biodiversity?[11]

This lack of clarity allows anything to be defined as 'sustainable' or as 'promoting sustainable development'. Therefore, there is a danger that the term will become a 'cliché, a fashionable phrase that everyone pays homage to but nobody will care to define'.[12] Moreover, analytical methods and policy making principles are not well defined.[13] For the concept of sustainable development to be operationally useful it must be more than just an expression of social values or political preferences disguised in scientific language. Ideally it should be defined so that one could specify a set of measurable criteria such that everyone in the decision-making loop could agree whether or not the criteria are being met in any given development programme.

Cities are fundamental to sustainable development.[14] They allow people to benefit from the concentration of economic activities and social interactions, connected as they are via a range of transport services, and in the process consume elements of the environment.[15] The environmental consumption and resource use per person is generally low in a high-density city in comparison to a low-density city.[16] However, people in Pakistan are moving out of high-density cities to the suburbs and from suburbs to the countryside largely because of a hostile, unclean, congested and polluted environment. As a result, a large proportion of urban development activities are now occurring on the peripheries of Pakistani cities.

Sustainable urban development is dependent upon the city being a centre of economic and social opportunity through efficient use of space.[17] In light of urban development activities over the last decade, travel modes in Pakistani cities are increasingly changing towards private transport, either motorcycles or cars. In addition to the growth in the number of private vehicles, this development has also resulted in increasing the length of each trip and enhancing car dependence which makes travel patterns more complex and dispersed. Development around car accessible locations is seen as one of the major limitations to the achievement of sustainable urban development.

THE ENVIRONMENTAL IMPACT OF URBAN TRANSPORT

The impact on environment due to current urban transport patterns is rising around the world, particularly in Pakistan.

Global Warming

Even though there may be disagreement about the abstract concept, in more concrete terms there is at least one requirement in the concept of sustainable development that has widespread agreement: greenhouse gas (GHG) emissions from all sectors of industry must be reduced. The science of climate change has explained how the earth warmed and cooled naturally at intervals over its lifespan of billions of years. The process results from the 'carbon cycle'[18] in which carbon in the form of gas is transferred between the main storages of carbon that include the atmosphere, the land biosphere and the oceans usually called 'carbon pools' or 'carbon sinks'. The flow of carbon into the atmosphere has developed a balance equalled by its reabsorption by the ocean and land biosphere over billions of years. What is new and dangerous about the current phenomena is the extraordinary rate of emission of carbon dioxide (CO_2) into the atmosphere from human activity, production of commodities and provision of services.[19] The emission of greenhouse gases is taking place largely because of the growth of the world's carbon-based economy over the last 200 years.[20] These emissions are disturbing the naturally developed carbon cycle which will cause more than a change in the weather.[21] Climate change will seriously affect every aspect of life on the planet, both human and non-human. Climate change is perhaps the first truly global problem of environmental sustainability that human society has ever had to face, and the world is ill- equipped to do so.[22]

Several reports of the Intergovernmental Panel on Climate Change (IPCC)[23] demonstrate with increasing evidence that the earth is warming as a result of human economic activities. The transport sector is a major source of CO_2 emissions, with urban transportation generating more than half the total. Moreover, greenhouse emissions in the transport sector are increasing more rapidly than in other sectors of industry.[24] Using a climate model, Wardle and Smith (2004) found that the GHG emission causes a complex response whereby some regions in Australia became wetter compared to other regions which experienced increasing threat of drought.[25] Urban transport has become one of the most significant activities which contributes to global warming and is thus responsible for defeating the objective of sustainable development. The link between unsustainable development and a certain, deep-seated policy behaviour that is highly resistant to change has emerged over a long period.

In the case of Pakistan, energy transformation and combustion activity is the most significant contributor to GHG emissions, accounting for about 70,000,000 tonnes, or approximately 60 per cent of total estimated emissions in the country.[26] More than 85 per cent of energy is produced using fossil fuels, primarily natural gas and petroleum. The share of the transportation sector in total energy consumption in the country is estimated at 32 per cent and the sector accounts for 23 per cent of GHG emissions. GHG emissions from the sector are estimated to be substantial as road transportation accounts for 90 per cent of passenger traffic and 96 per cent of freight traffic, relying almost exclusively on petroleum products.[27] It is evident that road transport is the largest energy consuming sub-sector in the transportation sector and the second largest GHG emitting sector after production industries. In spite of these figures, Pakistan's contribution to the global problem of climate change is relatively small: 94.5 million tonnes of CO_2 produced per annum in 2000, accounting for about 0.4 per cent of the world total.[28] Pakistan has a relatively low level of per capita GHG emissions with a total of 0.7 tonnes per capita compared with USA (19.7), Australia (17), Canada (13.8), Germany (10.5), UK (9.5) and France (6.2) tonnes of emissions per capita.[29] However, the implications of the polluted environment could be potentially much larger in future.

In order to address this issue, Pakistan signed the United Nations Framework Convention on Climate Change (UNFCCC) in 1992 and ratified it in 1994. The Country Case Study on Climate Change Impacts and Adaptation Assessments in Pakistan was completed in 1998 under the Asia Least-cost Greenhouse Gas Abatement Strategy (ALGAS). The Country Case Study report assessed the impact of climate change on major sectors of the economy, i.e. energy, agriculture and forestry. The transport sector has been discussed in the energy section of the report.[30]

However, the full impact of transport and related infrastructure policies was ignored. The policy focus relied on technical solutions in the road transport sector such as fuel efficiency which would have only a limited impact on GHG emissions while the overall level of emissions would continue to grow with increased use of fossil fuels.

Peak Oil

Fossil fuels, especially oil, are crucial to modern transport systems. Oil provides nearly 90 per cent of the world's transportation needs.[31] The modern transport system is built on ever increasing supplies of easily accessible and cheap oil.[32] As a result, world transport systems have become dependent on oil. However, it is generally agreed that oil is a finite resource and various estimates indicate that the peak oil production has already been reached or will be reached in the near future. 'Conventional oil' refers to the relatively easily accessible oil from the ground and does not include shale oil, oil from tar sands and deep water oil.[33] In this scenario, 'peak oil' means that the volume of oil production plateaus and then declines. Since supply will no longer be able to meet rising global demand the likely result will be high prices, potentially generating serious economic, social and environmental effects.[34]

The peak oil phenomenon was first observed in the US by petro-geologist M. King Hubbert in 1956. Hubbert observed that in any large region, unrestrained extraction of a finite resource rises along a bell-shaped curve, now known as the 'Hubbert Curve'; this curve peaks when about half the resource has been used up. To demonstrate his theory, Hubbert fitted a bell-curve to production statistics and projected that crude oil production in the lower forty US States would peak in 1969. The passage of time has confirmed Hubbert's analysis when production peaked in 1970 and continued to follow the Hubbert Curve with minor deviations.[35]

The timing of the peak oil production in the world is subject to uncertainty due to a range of factors such as estimates of ultimate recoverable reserves, increase or decrease in demand, efficiency gains, energy conservation and substitution by other energy sources.[36]

The Association for the Study of Peak Oil and Gas (ASPO)[37] has predicted a peak year of 2006 or 2008 on the basis of recoverable reserves of oil in the world.[38] The ASPO has estimated that reserves of 200 billion barrels of oil are expected to be discovered in the next thirty years. However, it is forecast that consumption of oil will increase to over 1000 billion barrels during the same period. The analysis of Campbell and Laherrere affirmed that production of conventional oil will peak during the first decade of the twenty-first century.[39]

The United States Geological Services (USGS) has produced three estimates based on different probabilities of finding additional recoverable oil resources, giving peak years of 2026, 2037 and 2047.[40] The main argument behind extending the peak year lies in a belief that there is a lot more oil yet to be found and new technology will make it possible to develop unconventional or alternative sources of oil.[41]

A report by the French ministry of economics, industry and finance on the global oil industry forecasts a possible peak in world production as early as 2013.[42] The 2013 peak oil prediction was forecast on the assumptions that oil demand will increase 3 per cent annually and that 10 billion barrels of new oil reserves will be found every year.

However, Heinberg argues that twenty-four of forty-four significant oil exporting nations have currently passed their production peak.[43] Campbell explained in his study that the global peak of the discovery of conventional oil reserves passed in the mid-1960s.[44] Thompson showed that a few major discoveries only result in an increase in production for a few years and then the downward trend continues.[45] McKillop argued that fast growing economies such as China, India and other East Asian countries put pressure on the demand for oil, which will bring the global peak sooner.[46] Wesolowsky found that oil-industry investment patterns seeking to maximize financial returns indicate that there is not much oil left to discover.[47] In 2004, the *Financial Times* quoted a study by the Scottish energy consultant, Wood Mackenzie, which showed that major oil companies had invested US$35–50 billion between 1998 and 2003 to develop existing oilfields. During the same period spending on oil exploration dropped from US$11 billion to US$8 billion. A similar trend has been found by Campbell[48] who points out that oil companies such as Shell have changed their focus from oil exploration to developing renewable energy while British Petroleum (BP) redefined its core business with the slogan 'beyond petroleum'.

In short, a key argument of the peak oil debate is that the world is not running out of oil, but the end of abundant and cheap oil on which the world depends is rapidly approaching. Since cheap transport is a key factor in economic growth, a rising oil price will have a long lasting affect on the global economy.[49] The peak will be a turning point for the world,[50] or the peak threatens a 'collapse' of modern society and its transport systems.[51] As the peak oil phenomenon will affect the transport system of every country in terms of its operating and infrastructure cost it is important to consider this debate in the context of Pakistan's oil consumption for transport.

Historically, Pakistan is dependent on oil imports. Pakistan produced 45.5 per cent (23,485,000 barrels) of crude oil locally and imported another 54.5 per cent (28,178,000 barrels) in 1990–91.[52] With time the dependence on imported oil has increased as Pakistan imported 71.8 per cent (57,699,000 barrels) of oil in 2003–4.[53] Moreover, one-third (34.7 per cent) of petroleum products was imported in 2003–4.[54] Furthermore, the local production of oil is forecast to start declining due to depletion of resources from the year 2010.[55] Crude oil imports for the year 2003–4 accounted for an equivalent of US$1801 million and that of petroleum products import was an equivalent of US$1318 million. The total annual import bill for the year 2003–4 was US$3119 million.[56] This increasing dependence on imported oil places an additional burden on the country's economy.

The transport sector is the largest consumer of petroleum products in Pakistan. According to the Economic Survey of Pakistan (2004–5) and the Hydrocarbon Development Institute of Pakistan, the transport sector accounted for an average of 48.7 per cent of total oil consumption between 1990 and 2004 followed by power generation (31.1 per cent) and industrial production (12.2 per cent) sectors.[57] During the same time period, oil demand from the transport sector increased by an average of 4.9 per cent annually.[58] To address the issue of an increased oil demand from the transport sector, the government launched the Fuel Efficiency in Road Transport Sector Project (FERTS) with the help of international donors. The objectives of the project were to improve fuel efficiency in the road transport sector by improved engine maintenance, vehicle maintenance, engine design and setting up of tune-up stations.[59]

The prices of petroleum products have been deregulated in Pakistan since 1 July 2001. The Oil Companies Advisory Committee (OCAC), a grouping of petroleum refineries and distribution companies in the country, reviews and fixes the prices of petroleum products every two weeks based on the international price benchmarks. The consumer price of all petroleum products includes government taxes. Since private and public passenger transport and freight movement are heavily dependent on oil, the cost of transport is directly linked to the traditional oil price.[60]

From the last decade onward, Pakistan has been encouraging the use of compressed natural gas (CNG) in automotive vehicles to reduce pressure on petroleum imports and to improve the environment. Pakistan is the third largest user of CNG in the world after Argentina and Brazil.[61] The Medium Term Development Framework (2005–10) has proposed that reliance on the import of oil be reduced by encouraging the use of CNG in private and public transport. However, gas consumption for the transport sector, mainly CNG, was only 1.5 per cent of total gas consumption in 2003–4.[62] Therefore, in the near future Pakistan will have to rely mainly on oil.

In summary, the transport sector of Pakistan is highly dependent on oil. Rising oil prices internationally due to the global 'peak oil' will have a long lasting impact on Pakistan's transport systems.

Urban Environmental Quality

Urban transport also generates environmental problems that are of a more local character. The most significant one is the increase of urban air and noise pollution which has become alarming due to the sharply growing number of cars. For example, the number of cars on the roads in Pakistan increased from 0.43 million in 1991 to 1.5 million in 2005, with an average growth rate of 8.7 per cent per annum.[63] In spite of these trends, Pakistani cities are still less motorized than other cities in the world (see Table 1.1). However, current motorization is mainly responsible for growing urban pollution. For example, surveys carried out in the country by the Pakistan Environmental Protection Agency (Pak-EPA) and the Japan International Cooperation Agency (JICA) in 2000 revealed the presence of an average of 895 micrograms per cubic metre (ug/m^3) of suspended particulate matter (SPM) on several road intersections of Lahore. These figures are 7.5 times higher than the World Health Organization (WHO) guidelines of 120 ug/m^3.[64] Similarly carbon monoxide (CO) levels in Lahore were found ten times higher than the WHO standard of 9 ppm.[65] Comparable figures have been observed for nitrogen oxide (NO_2) and sulphur dioxide (SO_2) in Lahore.[66] Studies carried out by the Pakistan Medical and Dental Association and the Aga Khan Medical Hospital have confirmed the presence of high concentrations of lead in the blood of school children, traffic police and adults in Lahore and Karachi. The presence of such high levels of suspended particulate matter and lead in the air is certainly a matter of concern because of its serious health implications for the public.

TABLE 1.1
CAR OWNERSHIP AND USAGE IN DIFFERENT CITIES

Cities	Car Ownership (Car per 1000 population)	Modal Split for Journey to Work Trips (in percentage)		
		Private Transport	Public Transport	Non-motorized Transport
Developing Countries Cities				
Lahore*	39	24	16	60
Karachi**	47	27	23	50
Mumbai***	21	9.3	40.9	49.8
Bangkok	199	60	30	10
Jakarta	75	41.4	36.3	22.3
Kula Lumpur	170	57.6	25.5	16.9
Developed Countries Cities				
Singapore	101	21.8	56	22.2
Hong Kong	43	9.1	74	16.9
Melbourne	518	79.4	15.9	4.7
Perth	523	86.2	9.7	4.1
Toronto	606	64.6	30.1	5.3
Vancouver	564	81.9	12.4	5.7
Zurich	444	36	39.8	24.2

Source: Kenworthy et al. (1999)
Note: Kenworthy et al. 1999 data refers to 1990 figures

* Lahore—Car ownership data was calculated from Census (1998) and Shigeta (2002) and refers to 1998 figures; Modal split data was obtained from TEPA and JICA (1992) and refers to modal split for total trips in 1991.

** Karachi—Car ownership data was calculated from Census (1998) and Malik (2006) and refers to 2006 figures; Modal split data was obtained from Malik (2004) and refers to 1990 figures.

*** Mumbai—Car ownership and modal split data was obtained from Low and Banerjee-Guna (2003) and refers to 1995 figure; modal split data refers to total trips.

Road Crashes

Urban transport also generates health problems more directly at the local level. The most serious problem is the growing rate of road crashes, which cause immediate and serious health, welfare, economic and emotional effects that are likely to exceed the indirect damage to health due to air pollution.[67] Currently, road transport causes 1.17 million deaths per year or 3242 deaths every day around the globe and this figure is predicted to increase by 67 per cent by the year 2020.[68] The number of injuries are ten times greater than the above mentioned mortalities. Pakistan ranked very high in the world for life threatening traffic fatalities where over 5000 deaths per year were noted in 2004.[69] According to WHO data, China, USA and Brazil respectively ranked top in mortality caused by road crashes.[70] Additionally, the number of injuries in Pakistan is ten times greater than the above figures.[71] The loss to the national economy, additional to human life, constitutes about 1.5–2 per cent of GDP or approximately Rs 60 billion per year in Pakistan.[72] However, most economists simply ignore traffic crashes and do not view them as an environmental and public health externality.

The circumstances of road crashes have become more serious due to the rapid increase of urban population and motorization in the high density 'mega cities' of Karachi and Lahore. For example, the population of Karachi increased from 5.2 to 9.2 million from 1981 to 1998 at an average annual growth rate of 3.4 per cent.[73] During the same time period the population of Lahore increased from 2.9 to 5.1 million by experiencing an average annual growth rate of 3.3 per cent.[74] On the other hand motorization in Karachi and Lahore increased with an annual growth rate of 6 and 6.3 per cent respectively.[75] The huge investment on roads to relieve congestion and for efficient movement of vehicles allowed people to adopt higher speeds, causing more serious crashes. The WHO studies explained that speed is estimated to be the main contributory factor in about half of all road crashes in low-income and middle-income countries.[76] For example an increase in average speed of 1 km per hour (km/hr) results in a 3 per cent higher risk of crashes.[77] In Lahore the number of fatal and non-fatal accidents increased by 11.4 and 10 per cent per year respectively between 1992 and 1996, corresponding to a significant improvement in road quality and layout.[78] The relationship between speed and crashes is particularly critical for vulnerable road users such as pedestrians and cyclists. For example, pedestrians have been shown to have a 90 per cent chance of survival when struck by a car travelling at 30 km/hr but less than a 50 per cent chance of surviving if the vehicle is travelling at 45 km/hr and no chance of surviving at 80 km/hr.[79] Worldwide data shows that two-thirds of the victims of road crashes are pedestrians of which one-third are children.[80] In developing countries, the injury rates of pedestrians and cyclists are much higher than in developed countries. In India, 95 per cent of those killed or critically injured in road accidents are pedestrians and cyclists.[81] Shanghai shows the same trends. Data from Karachi reveals that during the last five-years 46 per cent of total casualties were pedestrians.[82] Usually motorized and non-motorized road transport uses the same road space in developing countries. Vehicles constantly changing their speed to accommodate slow traffic generate areas of high risk for crashes.[83] In summary, any response to road trauma, urban pollution, energy consumption, climate change and efforts to meet local and international commitments must address the need to make changes in transport and infrastructure policies in Pakistan. These changes must prioritize modes of transport which have less impact on urban pollution, energy consumption and climate change. Whitelegg ranked non-motorized and public transport as sustainable modes of transportation based on their impact on the environment and their ability to help attain sustainable development.[84]

The common denominator in all these environmental issues—growing greenhouse gas emissions, rising oil consumption, deteriorating urban environment and escalating road crashes—can be found in public policy which is driven by the belief that the use of automobiles, the growth of motorization and continuous investment in roads will bring economic growth. In fact, public policy in this regard is influenced by technical approaches, guided by international institutions for improving the standard of living, resulting in an auto-oriented modernization strategy in Pakistan. Urban public policy is closely linked to material interests, available technologies and the values of government organizations.

PUBLIC POLICY AND PAKISTAN'S POLICY RESPONSE TO TRANSPORT AND THE ENVIRONMENT

Public policy making by government is one of the central processes through which communities respond to major social, economic and environmental problems. Policies emerge from identifiable patterns of interdependence amongst key social actors or policy institutions, and it is the character of these systematic inter-dependencies that shape behaviour, problems and solutions. A standard definition of public policy is an action that employs governmental authority to commit resources in support of a preferred value.[85] According to Considine, 'Public policy is the continuing work done by groups of policy actors who use available public institutions to articulate and express the things they value.'[86]

Policy-making is therefore a shared or group activity, even though the ultimate outcome often appears to be the preference of specific institutions or political leaders. In practice the financial resources and values in public policies are part of a 'policy system' spreading throughout the national and even the international arena.[87] The 'policy system' is the complex pattern of relationships between institutions, groups and networks based on shared understandings, values, common sources of agreement and disagreement, and patterned interactions.[88] Public policy is formulated within the policy system according to the evaluation of issues, problems and solutions by key actors.

PAKISTAN'S PUBLIC POLICY ON TRANSPORT AND THE ENVIRONMENT

Pakistan was one of the pioneering countries in the developing world which instituted deliberate public policies on transport. These policies were formulated in the form of Five-Year Development Plans and 'National Transport Policy' documents. However, the first comprehensive effort linking transport with the environment was the Pakistan Environmental Protection Ordinance (PEPO) followed by National Conservation Strategy (NCS) and the Pakistan Environmental Protection Act (PEPA).

Vigar argues that a large number of countries have dealt with environment and transport problems by building more roads.[89] That is also the case in Pakistan, where road building has been the main policy solution to tackle urban transport problems characterized by traffic congestion. Road building is justified just as it is done in the UK where people see congestion as the biggest single problem in transport.[90] In this policy, congestion is addressed by the 'predict and provide' method, where demand for travel by various modes is extrapolated and then an attempt is made to match the infrastructure to the potential demand. Similar models are used for planning public transport. Although no one denies the importance of relieving congestion, road building is not an appropriate solution as it is damaging for the environment and public health.

However, a road building policy has been interpreted as meeting the need for efficient road links for economic development. Therefore, one primary aim of every Pakistani government has been to facilitate the rapid and free movement of vehicles by constructing more roads in urban areas. Thus, although the level of road building proposed by federal and provincial governments has fluctuated in the past, the principal transport concern of government has been to roll out road construction programmes. In addition, the government constantly encourages the automobile industry and the use of automobiles by instituting a number of measures like introducing a leasing policy and reducing taxes on the purchase and import of cars. The government has acknowledged that global pressure[91] is the main reason for adopting these policies.[92] But in contrast, the government is considering the introduction of 'road user's charges' for all modes of transport without taking into account the fact that poor people depend on public transport.[93] Government policy has resulted in a 50 and 100 per cent growth of car and motor cycle manufacturing in the last three years. The government believes that the automobile industry is playing an important role in Pakistan's economy; a foreign investment of Rs 35 billion is expected in this sector,[94] which would result in the creation of more jobs and a reduction in the price of cars.

In practice, the politics of urban transport and environment involves huge infrastructure projects in the big cities of Pakistan. These investment projects are advocated by studies conducted by foreign consultants and international development institutions who are prepared to invest in this sector. For example, in Karachi, all urban transport solutions lie in proposed infrastructure projects like the Lyari expressway, the Karachi Circular Railway (KCR) and a rail mass transit system which could be either the standard underground system or the more modern elevated railway (mono or magnet trains). However, the recent introduction of CNG buses and enthusiasm for the rehabilitation of KCR will help improve the environmental situation of Karachi. In addition, the government of Sindh is also seeking foreign investors (Sweden) to construct an automobile plant for buses.[95] The situation is the same in Lahore where construction of the Light Rail Transit (LRT), the Ring Road and the Elevated Expressway cover all discussion on urban transport—ignoring their huge cost[96] and the country's weak economy. The government is also looking for foreign donors to introduce franchised bus services in five major cities of the Punjab in order to improve the image of public transport.[97] The elimination of *tongas* is also an important government policy to improve the image of road traffic and reduce road congestion. However, this policy has encouraged the introduction of the 'motor-cycle rickshaw', which is a greater threat to the local environment in terms of air and noise pollution.

Every Pakistani government has been reluctant to address environmental issues in the transport sector because of the perception that dealing with them is incompatible with economic growth. Environment has been regarded as a luxury that cannot be afforded even in a time of greater economic prosperity. Although walking and cycling are economical transport modes and 60 per cent of non-motorized trips are less than 2 km in length in Lahore, no politician or transport professional has encouraged improvement in or provision of facilities for these modes.[98] Instead decision-makers pursue transport policies that work against walking and cycling. Public transport can provide high quality services for urban populations at much lower cost than a system devoted to private motorized transport and road expansion. Almost no attention is paid to the ways and means by which the demand for motorized transport could be reduced. The explanation for this neglect of the obvious and pursuit of policies that make things worse is the supremacy of the economic imperative, with a strong belief in the link between new roads, shorter journey times, inward investment and jobs.

The resulting investment in roads at the expense of the pedestrian environment has changed a pedestrian-friendly environment to an automobile-friendly one. In this situation, Pakistan needs to learn from the analysis of Barter that Asian cities which gave priority to public and non-motorized transport investment over private-transport oriented investment have achieved important long-term economic and social benefits.[99] On the other hand, the 'cars and roads' solution has become redundant in solving transport problems in these cities.

Current policies give some consideration to the environment. However, it is restricted at best to marginal 'environmental improvement' for specific projects. The government has made some effort to address the issue of vehicular pollution by introducing greater use of technology and by formulating appropriate laws. Lead-free fuel, conversion of petrol vehicles into CNG, establishing tune-up stations for petrol and diesel vehicles, implementing Fuel Efficiency in the Road Transport Project (FERTS) are some examples of these palliative policies. The government is also beautifying the landscape around roads and junctions in order to give the appearance of a clean environment. Some other measures taken by the government include the establishment of two environmental protection tribunals, instituting motor vehicle emission standards, reviewing and strengthening of the motor vehicle examination system and constituting environmental squads of traffic police. The most significant achievements in recent years are incorporation of environmental concerns in government policies and initiation of a process of environmental impact assessment (EIA) in all development schemes, especially in major road projects and new township development. However, because of weak institutions, lack of implementation and a low level of awareness, these measures have not made any significant impact. It is certainly true that the introduction of more technology such as fitting catalytic converters and incorporating EIA has led to a reduction of emissions. However, is it enough to make urban transport sustainable in Pakistan? Not if sustainability means more than local environmental improvement. Then the emphasis is so heavily on the minority mode of car transport, there can be little hope of even local environmental improvement for pedestrian traffic—which constitutes the majority of the public.

In summary, all efforts for sustainable development rely on a combination of a road building strategy to relieve congestion and technical solutions to solve local environmental problems. An initial examination of public policy reveals that current efforts are not fulfilling the criteria for sustainable transport: integrating economic, social, environmental and institutional issues; institutional issues get the least attention by technical experts involved in formulating these policies. All documents acknowledge that limited financial resources and lack of institutional capacity are the main barriers for the development of transport in Pakistan. But neither have any substantive proposals been made on urban transport planning for implementation by the local or provincial governments. The institutional perspective is important for developing countries where governments mainly depend on aid from international institutions which remain in the decision-making 'driving seat' of all major urban development investments, including transport investment. An in-depth analysis of institutions is necessary to explore the urban transport policy-making process.

INSTITUTIONS AND URBAN TRANSPORT

The 'policy system' is a complex pattern of relationships mainly amongst institutions based on shared understandings, values and patterned interactions. These patterns may take the form of 'political economy' or 'policy culture'.[100] The pattern of roles and resource flows in 'political economy' suggests a structure of power. However, analysis of this structure may not reveal the values and priorities engaged in policy making. Each policy system has a unique culture made up of characteristic values, preferences and habits of interaction. Policy systems contain the means by which issues, problems and solutions are evaluated as a part of a 'policy culture'.

To understand 'policy culture', Considine suggests five factors: (1) values: can be explored by investigating the practices and preferences of decision-makers; (2) assumptions: key assumptions are employed to support these values; (3) categories: policy cultures locate these kinds of values and assumptions within groups, allowing them to explore their preferred outcomes; (4) stories: cultural structures will often be communicated through 'common stories' conveying 'priorities'; and (5) language: in many cases, these basic building blocks—values, assumptions, categories and stories—are woven together into a characteristic policy language having its own vocabulary and grammar as part of action and expressed value structure.[101] All these factors are delivered through 'policy institutions'—the particular machinery of government that has evolved to control and authorize the use of resources by establishing priorities, fixing values and turning profound philosophical problems into the simple routines needed to sustain policy.[102] Immergut explained that public policy is not an outcome of the aggregation of individual preferences or political choice, technological processes and market forces or even of vested interests.[103] Institutions shape public policy from highly complex combinations of factors that include the features of political regimes and struggles for power. Institutions provide a context for action. If this context is studied carefully it would help us to understand why decision-makers make specific choices.

Since institutions play a key role in policy formulation, there is a need to define the term with greater precision. Institutions are the subject of research in many scientific disciplines, including economics, political and organizational sciences. There are various interpretations between and within these disciplines of the concept of 'institution', ranging from formal constructions such as legal regulations and organizations, to behavioural patterns such as habits and traditions.

According to the Oxford English Dictionary[104] an institution is 'an established law, custom, usage, practice, organization or other element in the political or social life of a people'. Buhrs and Aplin argue that, 'institution refers to relatively enduring rules (formal and informal), traditions, customs and routines that guide human behaviour'.[105] In the theory of 'Normative Institutionalism' in political science, the word 'institution' is used loosely to mean everything from formal structures to very amorphous entities.[106] For March and Olsen an institution is not necessarily a formal structure but rather is better understood as a collection of norms, rules, understandings and, perhaps most important, routines.[107] However, Kiser and Ostrom define 'institutions' as:

> ...rules used by individuals for determining who and what are included in decision situations, how information is structured, what actions can be taken and in what sequence, and how individual actions will be aggregated into collective decisions.[108]

Peter Hall argued that institutions were 'The formal rules, compliance procedures and standard operating procedures that structure the relationships between people in various units of the polity and economy'.[109]

Rather than focusing on formalized structures, this definition provides a sense of institutions as rules and procedures as described by Kiser and Ostrom. One of the most widely used theoretical definitions of institutions is that of Douglas North:

> Institutions are the rules of the game in a society or, more formally, are the humanly devised constraints that shape human interaction. They are made up of formal constraints (e.g. rules, laws, constitutions), informal constraints (e.g. norms of behavior, conventions, self-imposed codes of conduct), and their enforcement characteristics. Together they define the incentive structure of societies and specifically economies.[110]

North's definition can be interpreted as referring to an institution being a set of rules, but it is also possible and of considerable utility to allow a hierarchy of aggregation of individual rules. Thus, the set of official formal rules of a sport can be considered to be an institution. North specifically makes a distinction between the concepts of institutions and organizations, warning against confusing the rules of the game with the players. The purpose of the rules, he explains, 'is to define the way the game is played. But the objective of the team within that set of rules is to win the game—by a combination of skills, strategy, and coordination; by fair means and sometimes by foul means'.[111] Thus, teams (the organizations) and their strategies respond to the rule set, and are different in kind to the rules themselves (the institutions).[112] In a different analysis, the institutional approach uses a number of key tools, namely identification of key 'policy institutions', resources available to these institutions and policy discourses to investigate existing policy approaches and the interpretations of key policy issues.

The 'new institutionalism' in political science is now sufficiently developed to extract useful thoughts for an institutional approach. Hall and Taylor identify three schools of thought: 'historical', 'rational choice' and 'sociological institutionalism'.[113] For the latter Torfing uses the term 'social-constructivist institutionalism':[114] hence historical institutionalism (HI), rational choice institutionalism (RCI) and social constructive intuitionalism (SCI). The 'new institutionalism' is opposed to the older tradition of institutionalism in political science and pays more attention to the informal aspects of institutional life, particularly networks of actors and their cultural norms subsisting within and between formal hierarchies.[115] Moreover, new institutionalism pays more attention to the dynamics of institutional change.

Historical institutionalists, Hall and Taylor, define institutions as 'the formal or informal procedures, routines, norms and conventions embedded in the organizational structure of the polity or political economy'.[116] Generally, historical institutionalists associate institutions with organizations and the rules made by formal organizations. Historical institutionalism situates the analysis of institutions within the structural development of the state and mediates conflicts among powerful social actors so as to privilege some interests while demobilizing others.[117]

RCI attributes a somewhat larger role to the rational calculation of individuals and groups in the formation of institutions.[118] If there is a difference between HI and RCI it is that the former school is most interested in the way rules and conventions come into being while the latter focuses more on the choices made within the rules. The overlap occurs where new rules are being created within the old which is arguably how institutions generally evolve. Thus Torfing finds much common ground between them.[119]

SCI on the other hand introduces the additional aspect of discourse and culture into the picture of how institutions are maintained and how support for them is mobilized. SCI introduces into the definition of an institution 'symbol systems, cognitive scripts, and moral templates that provide the "frame of meaning" of action'.[120] Where RCI tends to view 'interests' in utilitarian terms of status, influence, money and security, SCI suggests that interests are constituted by a variety of more complex values structuring individual and organizational identity, many of which do not have to do with a narrowly inscribed 'self interest'.

In short, new institutionalism provides insights into what stabilizes institutionally entrenched behaviour, how institutions influence intentional action in the form of public policy, and what changes them. Thus policy is seen as a result of politics that takes place within an institutional regime.

The new focus of political science suggests that planners need to pay attention to the institutional obstructions in the way of desirable new policy directions as much as to new solutions. This is especially true of urban transport which has the potential to create large environmental problems both locally and at the global level. It is widely assumed that transport solutions in developing countries are primarily a matter of the export of technology from the developed to the developing world. This export trade has been going on for much longer than the current round of globalization driven by information technology and footloose capital organized by international finance. Nevertheless, it is unquestionably an accelerating feature of international institutions, affecting particularly the urban transport policies of developing countries.

Although technology has played an integral role in the development of all modern societies and is sometimes referred as 'the engine of economic growth', the challenge for developing countries is to adopt appropriate technology wisely and humanely. Defining appropriate technology depends on local needs and local social, environmental and economic circumstances, for no single technological solution can be assumed to be best for all situations around the world.[121] International institutions, however, increase the uniformity of urban transport solutions all over the world. So a taxi driver in Bangkok can proudly comment to a client: 'With all of our traffic, we have become modern, we are just like New York'.[122] Similar trends are observed in China, where non-motorized trips can still comprise up to 80 per cent of total daily trips, but transport priorities are moving to adopt Western trends of motorization.[123]

The situation is similar in Pakistan, where urban transport policy is neither strong enough to self-finance its development to meet the existing demand, nor do government organizations have the capacity to develop policy for the welfare of the largest part of the urban population. Therefore, government has to depend on international development institutions to support urban transport activities in large cities. It is widely assumed that the public of poor countries hanker after that form of development, and indeed 'development' is often defined as a high level of car ownership, capital-intensive road projects and high-tech public transport. Recent capital-intensive, consumption-promoting and more or less private-vehicle oriented transport policies in all Pakistani cities are the common feature of this development scenario.

The result of these globalized solutions has a potentially large impact on urban transport systems and the environment in Pakistan. These policies will change the modal split and stimulate growth in mobility thus having a direct impact on the environment and transport system in the near future.[124] These globally influenced infrastructure projects are often designed by the Urban Transport Planning Process (UTPP)[125] and justified by the principle of 'predict and provide'.[126] Yet they are unsuitable for developing countries because they are neglecting non-motorized transport and the current need for improved public transport. A central assumption of the UTPP is that people are willing to spend more to reduce travel time. This assumption is true for some travellers. However, in developing countries, large numbers of people prefer to spend more time on travel and pay less because they are already spending close to 25 per cent of their income on urban transport.[127] Although locally designed solutions for urban transport are appropriate, every city of Pakistan is under tremendous pressure from international agencies to adopt capital-intensive projects that emphasize investment and construction rather than improve operations and maintenance.

Historically, international institutions have played an important role in delivering financial and technical assistance to Pakistan. Such influence is enhanced by economic globalization. Burki argues that due to this assistance international models have started to impinge on national discretion.[128] Qadeer pointed out that every new development paradigm and every contemporary idea in the world has been imported under the public aegis in Pakistan because of international persuasion.[129]

The reasons behind the influence of international institutions are complex. The financial and technical assistance provided by international institutions give them the leverage to influence policies in the urban transport sector where they play an important role in infrastructure planning and policy. Lack of indigenous financial resources and technical support therefore emerge as significant factors responsible for the formulation of inappropriate urban transport strategies for Pakistan. In developing countries such as Indonesia, and certainly in Pakistan, local institutions also seek funds from international institutions. In many cases, priorities of the international development institutions to transfer technology conflict with sustainability priorities of urban transport as well as with some original objectives of these very institutions.

The role of international institutions and their relationship with national and sub-national institutions cannot be ignored if one is to map out decision-making processes for planning and development of urban transport in Pakistan. Currently the role of international institutions in the existing institutional structure of Pakistan is itself becoming a matter of serious public debate.

INTERNATIONAL, NATIONAL AND LOCAL INSTITUTIONS IN THE URBAN TRANSPORT CONTEXT OF PAKISTAN

Pakistan is a federal republic with three levels of government: federal (national), provincial (state) and city (local). Urban transport policies and programmes are the responsibility of the provincial government in collaboration with the city government. The federal government has no formal constitutional control over planning agencies or development activities in major cities and provinces. In practice, however, city and provincial governments depend on the federal government for funds to finance most of their transport programmes.[130] By virtue of this financial control, the federal government dominates the provincial and city governments.[131] The federal government collects 92 per cent of the gross revenues and the provincial and city governments collect only 8 per cent. This means that the revenue base of provincial and city governments is very limited and heavily dependent on federal government grants.[132] The provincial and city governments can enjoy the residual powers and impose any tax not reserved for the federal government. In terms of urban transport, the provincial government's main source of revenue is taxes on motor vehicles. City governments also get revenue from tolls on bridges and roads.[133] On the other hand, their public spending is mainly the subsidy provided to urban public transport.

Transport infrastructure in major cities of Pakistan requires substantial funding from the federal government. Traditionally, federal expenditure on transport infrastructure has overwhelmingly favoured roads. While the federal government provides a larger portion of funding, provision of transport infrastructure remains the responsibility of provincial and city governments who must then determine priorities in urban transport planning and infrastructure development to achieve their vision for their own cities. However, if the federal government earmarks money for roads, political factors make it difficult for provincial and city governments to divert this allocation to other projects.

The roots of this strong tradition of political and economic centralization can be traced back to the governance pattern in British India where the centre played a critical role in policy making.[134] Waterston argues that the political and economic centralization in Pakistan has been made possible because domestic funds and a large amount of foreign aid and foreign exchange resources are channelled through the federal government to provincial and city governments.[135] As a result, the federal government not only plays a major role in policy making but is also involved in all public projects and programmes, including transport, that require a substantial outlay of funds.

The Planning Commission, the national planning authority, prepares the country's Five-Year Development Plans and also plans for all mega projects including those for the transport infrastructure. These plans and projects are implemented after the approval of the National Economic Council (NEC) and the Executive Committee of the National Economic Council (ECNEC). The Planning Commission also prepares the federal annual development plan and reviews provincial development projects and their intra-project priorities.[136] A significant function of the Planning Commission is that it is in charge of the official development assistance from international and bilateral agencies.[137] By virtue of its control over domestic funds including the country's foreign exchange reserves, as also its control over the large amounts of foreign assistance, the federal government assumes a dominant position in the hierarchy of governments in Pakistan.[138]

Reviewing the amount of foreign borrowing and aid, Husain found that 80 to 90 per cent of development expenditure in Pakistan was met through foreign assistance.[139] Burki demonstrated that Pakistan's reliance on external flows to finance development projects increased over time.[140] International institutions therefore have had ample opportunity to play a significant role at all levels of development in Pakistan.

From the beginning, the Planning Commission has used foreign technical and financial assistance to support planning and development in Pakistan. For example, foreign advisors recruited and supervised by the Graduate School of Public Administration, Harvard University, were extensively involved in preparation of the First and Second Five-Year Plans from 1955 to 1965, with grants from the Ford Foundation. However, the pattern and level of involvement of international institutions providing technical assistance and funds has changed over time.[141]

Today, most international institutions are involved in the process in two different ways. In one case, Pakistan relies on the current practices of donor countries and international lending institutions which is to develop comprehensive plans or identify specific projects that can then form the basis for aid or loans. International institutions ask the Planning Commission to specify their funding requirement along with sectoral prioritization. It then becomes the responsibility of the Planning Commission to ask the planning and development (P&D) departments of provincial governments to identify their needs for assistance within the sectors nominated by the international development institutions.[142] The Planning Commission arranges loans from these institutions on the basis of proposals received from the provinces.[143]

The other process which allows involvement of international institutions in Pakistan is the methodology and governmental procedures that regulate preparation of provincial annual development plans, which include investment needs for urban transport. These plans are prepared at the provincial level but are largely based on the financial resources made available by the federal government. But some transport infrastructure projects require huge financial resources in which case the provincial government seeks help of the federal Planning Commission to not only make project presentations to foreign aid consortia but also in aid and loan negotiations. The Planning Commission organizes annual meetings of the Pakistan Development Forum—a consortia of foreign donor and lending agencies—during which it presents its needs for funding of various development projects, including those related to transport infrastructure. In this case, too, the Planning Commission has to accept the conditions of loans imposed by the international lending institutions.[144]

In both processes, international institutions require that comprehensive studies be conducted to evaluate problems, assess solutions and estimate the funding required for a project. But because Pakistani institutions, especially those at the provincial level, lack the requisite human and resource capacity, the federal government invariably ends up asking the donor/lending agency to have these studies conducted. Eventually the nature and magnitude of Pakistan's transport problem are determined by foreign consultants who are mandated to offer possible solutions.[145]

Due to lack of local knowledge and in line with the preferences, agendas and perceptions of international development institutions, these consultants and institutions are inclined to propose mainly road-based, capital-intensive, high technology public transport projects to relieve urban traffic congestion.[146] In Lahore, for example, the World Bank enriched a proposal for the ring road; JICA proposed a light-rail transit project. Similarly, for Rawalpindi/Islamabad the Swedish International Development Agency (SIDA) proposed a bypass road and a mass transit public transport system[147] in spite of the fact that Islamabad has been designed with a hierarchal network of wide roads.[148]

In most cases, these projects are justified by the UTPP.[149] Dimitriou found that in justifying investment decisions in land use and employment, the principal function of the UTPP in developing countries is almost similar to that in industrialized nations.[150] But this model ignores cultural differences, ethnic compositions, diversity of income and the large number of walking trips in developing countries. International development institutions and foreign consultants fail to consider these unique factors and recommend urban transport solutions adopted in developed countries for developing countries too. Dimitriou termed UTPP a homogeneous technique used all over the world.[151]

Dimitriou stressed that developing countries have many alternatives to make urban transport systems different from those in the developed world.[152] This analysis is based on figures which show that the majority of developing countries are less motorized, make a large number of non-motorized trips and have cities with high population densities. For example, data shows that vehicles per 1000 persons in Pakistan (23), India (34), Philippines (42) and Indonesia (81) are considerably lower in comparison to those in USA (779), Australia (616), Canada (585) and the UK (434).[153] Moreover, rising oil prices and oil depletion make it improbable that developing countries would ever be able to achieve the level of motorization that exists in developed countries. Dimitriou suggests that developing countries must determine the extent of mobility they require, and stop when that is achieved without over-designing their road network.[154] Developing countries have the advantage of being able to introduce a number of low cost measures such as public and non-motorized transport to address current transport problems, but on the other hand, compact and high density cities might take up huge investments to build limited access expressways in anticipation of future demands of motorization. Dimitriou argues that urban transport problems in developing countries can be solved by adopting indigenously designed development programmes that can be accomplished within their own resources or when necessary, by mobilizing additional resources.[155] However, such programmes, especially in urban transport, raise questions about the institutional and human resource capacity of local institutions.

The UTPP model has potentially a large impact on the urban transport systems and environment of Pakistan. Husain argues that if Pakistan wants the freedom to resolve by itself the country's development problems it should reduce dependence on foreign capital and on solutions proposed by international development institutions.[156] This issue lies at the core of transport planning and policy in Pakistan.

NOTES

1. WCED 1987.
2. Adams 1993; Sum and Hills 1998: 132.
3. Low 2003: 1.
4. Connor and Dovers 2002: 4.
5. Dovers 2001.
6. Lele 1991; Jacobs 1994; Dobson 1998.
7. Lele 1991: 613.
8. Ibid.
9. Jacobs 1994.
10. Jacobs 1999: 21–30.
11. Luke 1995: 21–22.
12. Lele 1991: 607.
13. Ibid.
14. Banister 2000: 115.
15. Low et al. 2005.
16. Newman and Kenworthy 1999.
17. Banister 2000: 117.
18. Hadley Centre, UK Metrological Office, http://www.met-office.gov.uk/research/hadeleycentre/models/carbon_cycle/index.html
19. Houghton et al. 1997.
20. Hadley Centre website, http://www.met-office.gov.uk/research/hadeleycentre
21. Low et al. 2005: 18–29.
22. Low 2003: 15.
23. http://www.ipcc.ch
24. Lidskog et al. 2003: 223.
25. Wardle and Smith 2004.
26. http://www.enercon.gov.pk/issues.asp
27. Govt. of Pakistan, Planning Commission 2005: Transport, 1.
28. UNDP 2000: 233.
29. Ibid., 231.
30. ADB 1998: 35–89.
31. Wesolowsky 2005 http://www.atimes.com/atimes/Global_Economy/GA26Dj04.html When Oil Peaks by Tony Wesolowsky in *Asia Times*, 26 January 2005
32. Griffiths 2005.
33. Smil 2003.
34. Griffiths 2005.
35. Campbell and Laherrere 1998: 81.
36. Griffiths 2005.

37. http://www.peakoil.net/
38. Wesolowsky 2005.
39. Campbell and Laherrere 1998: 81.
40. Oil Scenarios 2004 http://www.oilscenarios.info/index.htm41 Heinberg 2003.
42. Porter 2005; http://news.bbc.co.uk/1/hi/business/4077802.stm 'Peak Oil enters mainstream debate' by Adam Porter in BBC News, 10 June 2005.
43. Heinberg 2004.
44. Campbell 2003: 54.
45. Thompson 2004, http://www.wolfatthedoor.org.uk/'The Wolf at the Door, the Beginners Guide to Oil Depletion' by Paul Thompson.
46. McKillop 2004, http://www.petroleumworld.com/SunOPF112104.htm 'Oil price trends through 2004-2010' by McKillop, Andrew.
47. Wesolowsky 2005.
48. Campbell 2003: 65.
49. Low et al. 2005: 161.
50. Campbell 2003: 45.
51. Heinberg 2003; Deffeyes 2001.
52. Govt. of Pakistan, Ministry of Finance 2005: Table 13.2.
53. Ibid.
54. Ibid., Table 13.1.
55. Govt. of Pakistan, Planning Commission 2005: Energy.
56. http://www.mpnr.gov.pk/fosilfuel.php
57. Govt. of Pakistan, Ministry of Finance 2005: 189.
58. Figure derived from Govt. of Pakistan, Ministry of Finance 2005: Table 15.2, 189.
59. http://www.enercon.gov.pk/ferts/
60. Govt. of Pakistan, Ministry of Finance 2005: 86.
61. http://www.hdip.com.pk
62. http://www.mpnr.gov.pk/fosilfuel.php
63. Government of Pakistan, Ministry of Finance 2006: Table 13.4.
64. Ibid.
65. LDA 1997: 13–17.
66. LDA 2004a: 13–9; Shigeta 2002; Noman 2001.
67. Badami 2003: 13.
68. WHO 2004.
69. Federal Bureau of Statistics, www.statpak.gov.pk
70. WHO 2004.
71. www.enercon.gov.pk
72. Ibid.
73. www.statpak.gov.pk
74. WHO 2004.
75. World Bank 2006a; LDA 1997.
76. WHO 2004.
77. Ibid.
78. LDA 1997: 6–25.
79. WHO 2004.
80. Vasconcellos 2001: 196.
81. World Bank 1996: 65.
82. *Dawn*, 28 March 2002.
83. WHO 2004.
84. Whitelegg 1997.

85. Wikipedia Encyclopaedia, http://en.wikipedia.org/wiki/Public_policy
86. Considine 1994: 4.
87. Ibid., 6.
88. Ibid., 8.
89. Vigar 2002: 1.
90. DETR 2000.
91. In 2003, the federal finance minister acknowledged that due to European Union pressure, government reduced taxes on import of cars as well as taxes on the local manufacturers of automobiles such as Toyota, Honda, Yamaha, Nissan, Hino, Fiat, Massey Ferguson, Hyundai, and Suzuki.
92. *The News*, 9 June 2003.
93. *Daily Jang*, 10 June 2003.
94. *Nawa-i-waqt*, 14 May 2003.
95. *Daily Jang*, 26 June 2003.
96. *The Nation*, 23 January 2003.
97. *Nawa-i-waqt*, 10 June 2003.
98. LDA 1997: 6–7.
99. Barter et al. 2003: 269.
100. Considine 1994: 47.
101. Ibid., 14.
102. Ibid., 1994: 72.
103. Immergut 1998: 25.
104. www.oed.com
105. Buhrs and Aplin 1999: 318.
106. Robinson 1991 cited by Peters 1999: 28.
107. March and Olsen 1989: 21–6.
108. Kiser and Ostrom 1982: 179.
109. Peter Hall 1986: 7.
110. Douglass North 1990: 3.
111. North 1990: 5.
112. Connor and Dovers 2002: 7.
113. Hall and Taylor 1996.
114. Torfing 2001.
115. Ibid., 281.
116. Hall and Taylor 1996: 938.
117. Hall and Taylor, 1996; Torfing 2001.
118. Ostrom 1991: 238; Torfing 2001: 280.
119. Torfing 2001.
120. Hall and Taylor 1996: 947.
121. Roque 1996: 178.
122. *The New York Times*, 19 August 1994 cited by Cohen 1996: 25.
123. Smith 1997.
124. Rodenburg et al. 2002: 449.
125. Dimitriou 1992: 187.
126. Owens 1995.
127. Arif 2002: 371; Cohen 1996: 31.
128. Burki 2000: 86.
129. Qadeer 1983.
130. Waterston 1963: 27.
131. Ibid., 137.

132. Husain 1984: 105.
133. Ibid., 104.
134. Govt. of Pakistan, EUAD and IUCN 1992: 92.
135. Waterston 1963: 137.
136. LaPorte 1992: 108.
137. Ibid.
138. Waterston 1963: 137.
139. Husain 1984: 105.
140. Burki 1992: 129.
141. Ibid.
142. Transport and communication, energy, finance, agriculture and natural resources, industry and non-fuels minerals, social infrastructure and multi sector www.adb.org.
143. Burki 1992: 141.
144. Ibid.
145. Burki 1992.
146. Imran and Low 2005.
147. Govt. of Pakistan, NTRC 1995.
148. Frantzeskakis 1995.
149. Dimitriou 1992.
150. Ibid.
151. Ibid., 191.
152. Ibid., 214.
153. World Bank 2003: 5–6.
154. Dimitriou 1992.
155. Ibid.
156. Husain 1984: 115.

2

The Institutional Framework for Urban Transport Provision

Despite international and national commitments to make the transport pattern more sustainable, everything in Pakistan's transport policy tends to stay pretty much the same. Attempts to shift vision and policy, and alter behaviour often result in only marginal changes to transport policies which continue to be based on road building and increased mobility. Some efforts have been made to promote a sustainable transport policy but every new policy is often a continuation of previous policies.

This book aims to explain factors that contribute to resisting change in the pre-established transport pattern in Pakistan. To this end, perhaps the dominant challenge in urban policy analysis is to explore how one transport option is adopted above others in an institutional context. Institutions now stand at the heart of social sciences, and analyses from a variety of perspectives have produced compelling work on the origins of institutions as well as institutional change.[1]

> Contemporary social scientists typically take a 'snapshot' view of political life, but there is often a strong case to be made for shifting from snapshots to moving pictures. This means systematically situating particular moments (including the present) in a temporal sequence of events and processes stretching over extended periods. Placing politics in time can greatly enrich our understanding of complex social dynamics.[2]

The notion of 'path dependence' is one of the frameworks appropriate for conducting institutional analysis. In making this argument, institutional analysts follow the lead of renowned economic historians who claim that the analysis of path dependence opens whole new frontiers of investigation in economics.[3] The notion of path dependence is generally used to support a claim that 'history matters' or the 'past influences the future'.[4] The path dependence theory suggests that initial choices get 'locked-into' suboptimal policies enacted at a time when their future consequences are unclear. Is this the situation in Pakistan where initial choice of urban transport policy has become 'locked-into' suboptimal road-based policies which now resist change to a more sustainable policy framework? To answer this question, one must first develop a framework to investigate if path dependence might exist in the urban transport policies of Pakistan, and if so in what form. This chapter therefore proposes a theoretical framework derived principally from the theoretical analysis of the notion of path dependence.

URBAN TRANSPORT POLICY: OPERATIONALIZING A PATH DEPENDENCE APPROACH

Pakistan has prepared several policies to address environmental issues in the country but has not been successful in controlling greenhouse gas emissions, high energy consumption, deteriorating urban environment and rising road crashes. As has been noted, local institutions have formulated all urban transport policies in Pakistan which depend in turn on financial resources and technical advice from international development institutions. Together these sources create and put into practice the conventional discourse. The primary question is therefore: 'How do international development institutions influence governmental decisions at federal, provincial and local levels in Pakistan to incorporate sustainability objectives in urban transport?'

To answer this, I will set out a theoretical framework to probe and explain what can be observed. The use of a theoretical framework will help in developing more detailed research questions which can guide empirical study. This framework will also help to define the appropriate research design and appropriate data collection, and it also becomes the main vehicle for generating results.[5]

David and Arthur devised the concept of 'path dependence' which provides a strong basis to explain the process of how an outcome becomes 'selected' over time.[6] It can provide insights into institutional level decisions. One of the seminal articles on 'path dependence' is found in 'Clio and the Economics of QWERTY' by Paul David.[7] In this article, David attempts to explain how the typewriter keyboard became standardized even in the presence of alternatives that were more efficient. Further, the examples of different technologies such as the narrow gauge of British railways and the petrol engine provided fertile grounds for arguments based on the concept of path dependence.[8] This argument suggests that if alternatives to these products had over time been given an equal chance to establish themselves, the alternatives would have been better than those that eventually dominated. However, the dominant technology gathers a certain momentum because people derive increasing value to themselves from reproducing it. In short, an 'historical accident' in favour of QWERTY influenced the eventual outcome.

Path dependence means that 'the early history of market shares—in part the consequence of small events and chance circumstances—can determine which solution prevails'.[9] Arthur has applied these views to the evolution of the physical location of firms, and thus to patterns of urbanization:

> [We] cannot explain the observed pattern of cities by economic determinism alone without reference to chance events, coincidences, and circumstances in the past. And without knowledge of chance events, coincidences, and circumstances yet to come, we cannot predict with accuracy the shape of urban systems in the future.[10]

Brian Arthur's work concerns the sphere of production and the behaviour of firms.[11] He argued that self-reinforcing mechanisms exist in the logic of production to maintain the mass production of a given product. These are:

1. Large set-up or fixed costs: total cost per unit falls as output increases.
2. Learning effects: over time firms learn better ways of producing goods and at lower cost.
3. Coordination effects: producers look to what other producers are doing and match their behaviour to the behaviour of others.
4. Adaptive expectations: as a type of product increasingly prevails on the market, the belief grows that it will continue to prevail in future.

Douglas North has transferred the concept of path dependence from the sphere of material production to that of governance.[12] North argues that the competition is not amongst technologies;[13] rather it is amongst institutions that are making decisions. He explained that choices of policies are more multifaceted in an institutional context because of the complex interrelationships among formal and informal constraints. In consequence, path dependence appears much more complicated in the case of institutions than in the case of technology. His point is that the adoption of one technology rather than another depends mostly on institutional abilities and knowledge. North applied all features identified by Arthur as contributing to technological 'lock-in'—increasing returns and high fixed costs, learning effects, coordination effects and adaptive expectations.[14] He explained that in the context of complex social interdependence, new institutions often involve high start-up costs; institutions 'learn by doing' and get benefits from coordination with other institutions. Consequently, one could anticipate the same kinds of historical processes in which initial choice of institutional design has long-term implications for economic and political performance. In North's words:

> At every step along the way there are choices—political and economic—that provide…real alternatives. Path dependence is a way to narrow conceptually the choice set and link decision-making through time. It is not a story of inevitability in which the past neatly predicts the future.[15]

North interpreted path dependency as constraints on future choice sets. However, North's claim is not that path dependence freezes change for adopting future choices. He characterizes change as 'bounded' due to institutional continuity (Pierson 2000a: 76).[16]

Denzau and North separate the institutional field into two parts: 'Mental models are the internal representations that individual cognitive systems create to interpret the environment; the institutions are the external (to the mind) mechanisms individuals create to structure and order the environment.'[17]

Path dependence derives as much from the former 'shared mental models'[18] as from the external structure of rules and organizations. Institutionalized networks should be considered part of the external structure; common networks of ideas—shared mental models—are the internal counterpart. In this reference, Denzau and North discuss the process of discursive change.[19] In fact, Denzau and North explicitly acknowledge that 'many of the changes we wish to understand are social, such as changes in informal institutions or ideologies'.[20] From this perspective, policy documents may be seen to have a rather different function from that assumed by a conventional conception of rational politics.

For institutional analysis the concept of 'path dependence' provides a powerful metaphor for understanding and explaining institutional politics that might exist in transport decision-making in Pakistan.

WHAT MAKES A POLICY PATH DEPENDENT?

Arthur defined two primary components in the evolution of policies: (1) The period of selection, or the initial 'critical juncture', when events combine to develop a particular 'path' and (2) the 'reinforcing period' when supportive forces stabilize the 'path' according to the selection developed earlier.[21] These two primary components provide a theoretically grounded point of entry for exploring path dependent policy processes in urban settings. Let me elaborate each of these key features in detail.

'Historical Analysis' in Identifying Critical Juncture

There are rich traditions of historical analysis in the economic, social and political sciences in identifying 'critical juncture'. Arthur's phrase 'history matters'[22] or Pierson's concept 'history remembered'[23] are key concepts associated closely with the analysis of path dependence and can be described as 'evolutionary'.[24]

Arthur pointed out that the thesis of path dependence rests on a conception of 'historical causes' which are based on 'small events'.[25] For example, transport policies in the world mainly favoured road-based solutions on account of an historical event (it may have been due to the discovery of large amounts of oil, or to developments in manufacturing and engineering). However, if small events in history had been different, would the pattern of cities we have inherited be different in any significant way? Arthur said yes, and used the example of 'Silicon Valley' in the USA which became a high-tech city largely as the outcome of 'chance'.[26] He argued that a mixture of 'economic determinism' and 'historical chance' formed the spatial patterns of the 'Silicon Valley'. He concluded that we cannot explain the pattern of the cities by 'economic determinism' alone without 'historical small events', accidents and circumstances in the past.[27] In contrast, Peck argues that choice of product technology depends not on chance events but natural resources and capital.[28] He justified his argument on the basis of the early development of the motor car engine in the USA and Europe. On the same line, Woodlief argues that urban decision-making is 'purposive', that is, not based on the random selection of solutions.[29] He highlighted the issue of imperfect information, numerous distractions and little time that would make it possible for inferior solutions to gain favour at initial stages.[30]

Pierson argues that the concept 'history matters' is insufficient unless we are able to explore why, where and how.[31] He stressed that placing policies in time, systematically situating particular moments, including the present, in a temporal sequence of events and processes can greatly enrich our understanding of complex social dynamics.

Arthur stressed that 'timing' is crucial in explaining path dependence in the policy process, especially when decision-makers favour a particular solution.[32] Pierson explained that 'when' an event occurs is crucial in exploring path dependence in the policy process.[33]

Several authors also support the argument of 'sequence' to study path dependence.[34] David described path dependence as a type of process where 'one damn thing follows another'.[35] Pierson found that historical sequences provide essential building blocks in studying path dependence in the policy processes,[36] and argued that the 'temporal ordering of events or processes has a significant impact on outcomes'.[37]

Mahoney argues that a critical feature of path dependence in a policy process is the 'availability of alternatives' at early stages in a sequence.[38] Woodlief found that the path dependence process begins when choices of action existed in a sequence but a particular programme got priority.[39]

In conclusion, the theory of path dependence suggests that the present choice of policy is a reflection of prior events that occurred at a critical juncture, and that knowing those events can tell us why the present selection of policy and not other possible policies currently exist and are sustained. Our historical analysis will discover whether initial events locked policies into an inferior policy path. To understand how that path might continue to exist, the analysis needs to explore further self-reinforcing processes because by identifying these processes we are able to understand why institutional practices are so persistent.

Self-Reinforcement Mechanisms

The history, timing and sequence of events in the past in policy processes help us to understand the reasons for today's policies. The study of self-reinforcing mechanisms will help to explore factors and particular events responsible for stabilizing specific policies enacted at a critical juncture.

In the large emerging literature on path dependence, researchers have developed sets of features to explain self-reinforcing dynamics in technical production and institutional development. David argues that the 'locked-in' character of technology depends on compatibility of the 'hardware', that is the technology itself, and the 'software' which are skills people gain over time to run the technology.[40] The result of hardware and software compatibility would tend to reduce the cost, eliminate alternatives over time and ultimately become irreversible. For example, when through a 'historical accident', road transport becomes the dominant mode of transport, it leads to the development of hardware (the organizations of road transport) or HRD2, roads themselves (physical structure) or HRD3 and software (skills in engineering which support road use).

Pierson cited Jackson's example of post Second World War transportation policies in the United States[41] which encouraged massive investments in particular spatial patterns of work, consumption and residence. Once these policies had been implemented, their patterns sharply constrained the alternatives available to policy makers on issues ranging from energy policy to location of schools. These policies then encouraged individuals to develop particular skills (road engineering, automobile engineering), make certain kinds of investment (cars), develop certain kind of services (car and service infrastructure), and devote time and money to certain kinds of organizations (driving schools). In this context, policies became 'locked-in' to a particular path, and pushed individuals on to that path from which it is now difficult to extricate. Similarly, road and rail are to some degree alternative modes of transportation in most countries. Each mode is self-reinforcing in that the more heavily it is used, the more the funds that become available for investment in improvements which then attract more users. Therefore, one mode may achieve dominance at the expense of the other.[42] These examples illustrate the fact that incremental changes in technology, once begun on a particular track, may lead one technological solution to win out over another; even if that technological path is less efficient than the abandoned alternative, or may turn out later to be undesirable for other reasons.

North argued that all the features identified with technological lock-in are often also characteristics of institutions.[43] Mahoney has found that economic analysts employ a utilitarian theoretical framework—weighing the costs and benefits of policies—which is incomplete for studying institutional self-reinforcement.[44] For example, walking and cycling are the economical transport modes. A large proportion of trips in Pakistani cities even today are either non-motorized or on public transport. However, no politician or transport professional has encouraged provision of facilities for these modes. By contrast, decision-makers pursue transport policies that work against walking and cycling. Public transport can provide high quality services for urban populations at much lower overall cost than a system devoted to private motorized transport and road expansion. Although there is demand for road use from an affluent minority, yet almost no attention is paid to reducing the need for motorized transport. This shows that some other factors in parallel to rational decision-making exist which play an important role in contributing to the self-reinforcing mechanisms. Organizational theorists have developed a list of factors that contribute to resisting change in self-reinforcing policies; these include power, institutional decision-making structures and discourse.[45]

Power: Pierson argues that one of the key factors in the self-reinforcing mechanism of path dependency is 'power' in an institutional context.[46] Scott wrote that power is the capacity to influence others.[47] It is a social relation between actors or organizations in which one organization is dependent on another. Scott pointed out that sometimes even strong organizations have no powers in formalized relationships as when they are situated within a structure of power.[48] However they remain dominant by applying pressure which makes it possible for them to exercise influence in the decision-making process. Such pressure might be exerted by military victory, by international institutions in the context of development policies, by international obligations, or by economically and politically superior states.[49] The existence of pressures facilitates the spread of specific policies.[50] Gaventa demonstrated that the self-reinforcement mechanism over time simultaneously increases the power of a dominant organization.[51] In short, the key variable for exploring power relations amongst organizations involved in policy making is the degree and nature of their interdependence.

Structure of Institutions: North[52] believes that no single institution can work independently and in reality each is connected in 'the interdependent web of an institutional matrix'. Therefore, the action of one institution is highly dependent upon the actions of others. Whether one institution gets resources and support depends not just on what it does, but more on what other institutions, locally and globally, do. In the urban transport context, road based urban transport models spread from the USA to many other countries after the Second World War. But these models failed to solve the urban transport problems of developing countries.[53] The reason for this failure rests on two arguments: (1) The nature of transport problem in developing countries is different to that in developed countries and (2) the urban transport plan-making process, the planning machinery, the institutional capacities, the norms, behaviour and the subjective models of organizations are entirely different.[54] As a result, the consequences of policies are different. In short, a common set of rules will lead to widely divergent outcomes in societies with different institutional arrangements. Therefore there are strong grounds to believe that self-reinforcing processes will be widespread in 'the interdependent web of an institutional matrix'. These processes need to be explored.

Discourse: North has argued that actors or organizations operating in a complex institutional system are heavily biased in the way they filter information with their existing 'mental maps'.[55] They all interpret complex issues in accordance with their own ideological beliefs or preconceptions and hence determine different paths. Daly found the 'economic growth'[56] mental model in his analysis of 'World Development Report: Development and the Environment' by the World Bank in 1992. The illustration of Herman Daly has relevance for the Pakistan case study described below. Daly recalls that he tried to introduce the notion of environmental limits to the senior economists of the World Bank while he was himself an officer in the economics department of that institution.[57] Daly comments that the Bank would be the proper institution 'to recognize the ecological contradictions in the world's development plans' and that the Bank's 1992 'World Development Report' entitled 'Development and the Environment' represented the best opportunity, to that date, of doing so. However, he writes that in the end the message of the published report was 'both a reaffirmation of the Bank's faith in economic growth and a denial of the existence of any fundamental ecological limits to that growth'.[58] Daly tells how he tried unavailingly to alter a diagram in the text entitled 'The Relationship Between the Economy and the Environment' that consisted of 'a square labelled "economy" with an arrow coming in labelled "inputs" and an arrow going out labelled "outputs"—nothing more'. Daly wanted the diagram to depict the economy as a subsystem dependent upon the environment as a source of raw materials and a sink for waste. He suggested that this could best be shown by placing a larger box around 'The Economy' labelled 'The Environment'. At first the economists agreed to the box but not to the label, and Daly commented that without the label the box was 'merely decorative'. In the end Daly's suggestions were ignored altogether. Daly later attended a conference at the Smithsonian Institution discussing the book *Beyond the Limits*,[59] which happened to contain a diagram identical to the one Daly had proposed. The chief economist of the World Bank, Lawrence Summers, was also present, so Daly asked him whether he felt that the size of the economic subsystem relative to the total system was an important one for economists. Daly's question was, 'What is the optimal scale of the economy relative to the environment?' Daly relates: 'His reply was immediate and definite: "that's not the way to look at it"'. Here then is a case where learning had occurred in an institution—the Bank's senior personnel had learned that the environment and development were connected—but the 'model space' defined by the central assumption, that all problems can be solved by economic growth, had not changed. It may be that the World Bank's model space has subsequently changed, but that is something this study has yet to establish.

This discussion underlines the importance of belief of the policy makers in carrying out an institutional analysis. It is therefore important to explore the beliefs of those who are directly or indirectly involved with decision-making and how they filter information and promote information.

TYPES OF FIELDS IN FINDING PATH DEPENDENCE

The components of path dependence include the tools of historical analysis and the concept of self-reinforcement. Historical analysis helps in determining if there has been one or more critical junctures in policy history which initiated path dependence. If there was a critical juncture then the concept of self-reinforcement is employed to determine whether the current policy path remains unchanged, or is completely changed or modified from the policy formulated during that critical juncture in the policy history. The study of both components will help in identifying the nature of path dependence, e.g. complete, partial or no path dependence. The question then is how to measure the nature of path dependence in the policy outcome?

The current analysis might identify three conceptual fields to assess whether path dependence can be detected in existing transport policy in Pakistan. These are institutions, techniques and discourse. The first of these fields will explore whether institutional relationships, power and finance of national and international institutions influence the path of policy. The second field will determine if and what similarities there might exist in techniques used in the historical trajectory of the policy to produce a particular outcome. Finally, the discourse field will determine whether an internal network of ideas, together with their carriers, impact policy.

These fields of path dependence focus attention on the relationships, power and financial flows amongst different organizations; mode of technical methods in analysing and providing urban transport solutions and the frame of reference of decision-makers. These can be crucial issues in revealing the way organizations relate to one another for generating transport policies and solutions. The manner in which policies flow up from local level and obtaining resources from central organizations and international institutions, all relate to the techniques, institutions and discourse fields.

LOOKING FOR EVIDENCE OF PATH DEPENDENCE IN PAKISTAN'S TRANSPORT POLICY

It may be helpful to restate the primary research question 'How do international development institutions influence governmental decisions at the federal, provincial and local levels in Pakistan to incorporate sustainability objectives in urban transport?' into three sub-statements.

1. To assess whether path dependence can be detected in the transport decision-making process, the analysis needs to elucidate the institutions' field by considering how institutional relationships, power and finance influence local and international institutions.
2. To assess whether path dependence can be detected in transport planning and policy, the analysis needs to elucidate the techniques' field by considering which method was employed for analysing transport problems and generating transport solutions.
3. To assess whether path dependence can be detected in the perception of decision-makers, the analysis needs to elucidate the discourse field by considering how transport problems are defined in the minds of decision-makers, locally or internationally.

The theory of path dependence uses the organizing concept of historical juncture and self-reinforcing mechanism as a way of analysing policy. This book uses these concepts as tools to mobilize the analytical concern by using North's concept of 'interdependent web of institutions'[60] and Hajer's idea of 'storylines'[61] to explore whether path dependence can be detected in hardware (HRD2—institutions for transport planning and infrastructure provision) and software (perception and skills) of transport institutions.

Returning to the principle that 'history matters' the first aim is to conduct an historical analysis to explore whether any 'critical juncture' can be found in transport planning. Woodlief argued that it is important to look at the beginning of the path, when choices were made.[62] Pierson also stressed the need to explore 'when', beside 'what' happened in history, in institutional processes.[63] As it is not clear at this stage exactly when did the current transport policy begin, I will review transport development from the time of British colonial rule in India and document the type of transport policy adopted during that period. This period has also been selected because it was during British rule that modern transport (railway and roads) was introduced in the subcontinent. The discussion will continue by exploring the transport policy adopted in Pakistan after the British left. Documenting types of transport policies adopted over time will help in determining if there was a critical juncture that might explain the present transport policy of Pakistan.

Given that international institutions played an important role in transport development in Pakistan, the analysis must explore and then compare results produced by their involvement and that left behind by the British. Moreover, the analysis will identify when and why international institutions were involved in the transport sector in Pakistan. To answer the questions, the investigation needs to conduct a historical review of development plans after independence and chart the evolution of transport development in Pakistan. Contextual analysis will help to expand our understanding of the initial choice of transport development. A historical review will help to locate facts important to answer the questions of 'when' and 'why' raised earlier. Reviewing the history of institutional and organizational features of the established system will also help to locate these facts.

Secondly, the study should review the 'historical causes' which may have caused transport solutions to go along a particular path. Arthur revealed that economic argument should be studied in parallel with 'small historical events' to determine the pattern of urban policies.[64] On similar lines, Woodlief argued that urban decision-making is 'purposive'.[65] Therefore, it is important to explore 'how' transport decisions were initially justified so as to assess whether those techniques continue to be employed even today. If the current study uses Arthur's and Woodlief's argument in a transport context, it appears that road-based transport policies are perhaps favoured on account of the supremacy accorded to economic activities inherent in roads development. Therefore, this study would explore the economic justification or purpose in opting for the existing path of transport and to what level that has been achieved. This discussion needs detailed investigation to highlight Woodlief's argument of imperfect information, numerous distractions and little time for initial selection of the solutions.[66] According to March and Olsen the inferences of past experience are stored in standard operating procedures, professional rules, and the elementary rules of thumb of a practical person.[67] These components can be assessed by accessing institutional documents and organizational charts.

The second component, i.e., the self-reinforcing mechanism will also be explored in the current investigation to identify whether path dependence can be exhibited in the transport policies of Pakistan. Brian Arthur lists four features—large set up costs, learning effects, coordination effects and adaptive expectations—for studying self-reinforcing mechanism and these will be used to discover whether Pakistan transport policies are path dependent.[68]

The first argument is on the effect that the transport policy has on different institutions. Pierson argues that 'policies provide both incentives and resources that may facilitate or inhibit the formation or expansion of particular groups'.[69] Thus, initial policy encourages certain institutions to mobilize for the maintenance of the programme or its expansion. So this analysis must explore institutions that enjoy an advantage which is used to affect the policies of decision-makers. Pierson explained that institutions controlling resources such as money, information, skills and personnel affect government policies.[70] Therefore, the current investigation will identify those who have conducted transport studies in the past and have provided funds for these studies and programmes. This will help to explore dominant institutions and their control over important resources such as the technical and financial assistance needed for developing transport solutions.

A second type of policy mechanism involves bringing about policy change by improving overall institutional capacity. The viability of any policy rests on the capacity of an institution to develop and implement that policy. To assess institutional capacity implies exploring skills, technical knowledge, management systems and belief in existing institutions. This data should tell us if local and international institutions weigh heavily in the favour of the *status quo*.

A third mechanism which exists relates to policies that may have major consequences for individuals. Pierson states: 'policies may encourage individuals to develop particular skills, make certain kinds of investments, purchase certain kinds of goods, or devote time and money to certain kinds of organizations'.[71] However, existence of the individuals 'mental map' affects his or her choices.[72] Mental maps have been developed through 'storylines' promoted by policy actors.[73] Belief in certain storylines influences an individual's choice. Hence the study will explore different storylines which influence individual behaviour. In exploring self-reinforcing mechanisms in transport policy in Pakistan, the analysis will provide an account of whether the initial policy structure of transport is entrenched and has become resistant to change.

Historical sociologists such as Thelen and Mahoney enrich the study of reinforcing mechanisms by identifying additional factors including power, function (structure of institutions) and legitimization (law and resources).[74] SCI and organizational theory pays more attention to the discursive structure of institutions contained in policy documents revealing belief and language, norms and rules.[75] Understanding these mechanisms is important to find path dependence in the policy processes in Pakistan.

a) The Power of International Institutions

One of the objectives of this book is to explore reasons for the involvement of international institutions in the transport sector of Pakistan. Tentatively, it may be assumed that this is due to lack of financial and technical capacity in government institutions. However, for detailed analysis, it is important to explore exactly when these international institutions started to get involved and precisely at what level do they interact in the existing institutional framework of the country to organize decisions.

There is therefore a need to conduct a historical review of the country's development plans following its independence. This will help to investigate the role of international institutions over time in areas beyond the formal structure of government in Pakistan. More specifically, reviewing early development plans and the history of current institutional and organizational features of the established system will help locate these facts.

The study requires mapping the pattern of involvement of international institutions at different levels of government, which in turn requires drawing a chart to show how they organize decisions in the country's transport policy, and at which level. In the current study, I am more interested in exploring the relationships between national and international institutions because at certain times the national government works as an agent of international institutions and reduces the capacity for indigenous solutions.

Institutional analysis based on the above framework provides a way of looking at the complex policy issue of 'power'. It will explore whether power and authority is in the hands of international institutions by highlighting financial and technical dependencies of existing institutions in Pakistan. Finally, this will help to explore how decisions are made and who has power and authority to decide transport policies in Pakistan.

Investigating financial and technical dependencies will help explain where power lies and whether transport programmes are path dependent. This is done by evaluating the financial priorities of institutions and their criteria for delivering transport solutions within the hierarchal structure of government and international institutions working in Pakistan. The budgetary process offers a lens through which to distinguish underlying values—political, economic and social—in so far as they are reflected in specific decisions on raising revenue and distributing expenditure. Cash values reflect political values.[76] Treasury assumes a critical role in the assessment of institutional priorities, and the distinction between policy and resource issues fades. Though Pakistan has a high level of corruption and was ranked seventeenth from amongst 163 countries in 2006,[77] there is little possibility of corruption at the policy-making stage because that is a practice which would normally occur in the implementation of a development project. Institutional priorities can be assessed by studying budgetary documents to determine financial allocation in different years for road building, railways, public transport, non-motorized transport, research and institutional capacity building. Identifying criteria for funding from international institutions will reveal the values they espouse and their favoured methods for adopting different programmes.

b) The Structure of Policy Making Institutions

To find out how decisions are made in transport policy and planning in Pakistan, the study should focus on North's 'interdependent web of an institutional matrix'.[78] According to Considine nothing is more important in an institutional analysis than exploring the authority of 'policy institutions' by a closer scrutiny of relationships amongst various levels of government.[79]

Pakistan has three different levels of government: federal (central), provincial and local. All three governments have a 'web of formal agencies' bound together for transport issues. This study is interested to explore how these levels of government work together to make transport policy decisions and to this end will depict their linkages in a chart form. This study will also display in graphical form the dependency relationship amongst various formal organizations in the urban transport sector in order to identify the capacity levels of government in Pakistan. These capacities can be broadly defined in terms of substantive capacities (policy formulation, institutions, human resources, technological and financial) and support capacities (programme management and coordination, information management and budgetary).[80] Hence, the aim of institutional analysis is to examine the nature of institutions and the consequences of institutional dependencies for the sustainability of transport sectors.

c) The Discourse of Critical Assumptions and Beliefs

Policy discourses are a specific ensemble of ideas, concepts and categories that are produced, reproduced and transformed in a particular set of practices. Hajer argues that 'policy discourse' is the prime vehicle for change.[81] Policy discourse is influential on the actions of government for defining and promoting transport policies. The important question here is how will the current study explore 'policy discourse'? How do perceptions or discourses of different individuals or organizations influence one another? To answer this, literature has identified the following important areas of study.

Beliefs and Storylines

Gusfield argues that a public problem can be related to specific 'problem owners' who successfully impose their definition of the problem on others.[82] Dimitriou posits that organizations responsible for decision-making tend to hold different perceptions of what the problem 'really' is.[83] Arthur and North argue that actors who operate in the social context of a complex institutional system are heavily biased in the way they filter information into their existing 'mental maps'.[84] These perceptions or ideological beliefs vary with the analyst's interest, institutional allegiance, political ideology, education, training, background and cultural norms. Ideological beliefs influence the subjective construction of the models that determine different paths. These paths and policies are supported by organizations through investments in specialized skills that support existing institutional arrangements and belief systems.

Beliefs can be measured by identifying 'storylines' that stabilize different paths. Storylines are the linguistic representations of issues that provide ways of holding ideas together and enable arguments to be transmitted amongst stakeholders or key institutions.[85] The influence of the policy depends on the cognitive power of its storylines. Storylines are developed and coordinated through 'discourse coalitions'; such coalitions amongst actors are in turn formed with the help of a series of storylines.[86]

Hajer identified 'storylines' in his analysis of the influence of ecological modernization in regulating the problem of acid rain in Great Britain and the Netherlands.[87] According to Hajer's framework, discourse, storylines and discourse coalitions represent the perception of a problem, promotion style and the way actors join together to exercise power.[88] Vigar's analysis of the transport policy in the UK accords with Hajer's method.[89] His work reveals the existence of a collection of storylines such as 'local economic competitiveness', 'balanced approach' and 'heroic engineering'. Vigar's work shows how the 'predict and provide' paradigm found support from 'local economic competitiveness' storylines in the presence of demand management and sustainability rhetoric in transport policy.[90] He uses the term 'balanced approach' to show how transport investment in the 1960s was promoted by increasing the capacity of public and private transport networks without incorporating demand management.[91] Similarly, he explained that the term 'heroic engineering' was used to describe a particular period of transport planning in the 1960s.[92] These storylines are also found in the work of Mazza and Rydin comparing traffic policies of the UK with Italian cities.[93] The above discussion suggests that this study should explore storylines in order to learn how problems are perceived and promoted in particular policy agendas.

Policy papers are an important source of 'storylines', even if they do not include 'hard' new proposals or legislation. It becomes imperative to examine the specific conception of the reality or '*status quo*' as something that is upheld by key actors through discourse. Along with policy papers, policy analysis also requires examination of current practices and programmes, legislations and use of metaphor. The manner in which environmental issues are addressed in the case study of a project, the contents of the Environmental Impact Assessment (EIA) and the method adopted in conducting the EIA will all be reviewed in order to analyse the perceptions of government in relation to transport projects. This will help to explore how sustainability is defined and how storylines are promoted in a case study.

Hajer's theoretical concepts are a valuable source of inspiration. As transport policies involve a slow and long process of planning and practice, it is useful to locate storylines historically, which will enable us to chart the manner in which transport discourses were constructed and contested at different times in Pakistan. Identifying these storylines will help us to understand how problems are perceived and promoted in particular policy agendas.

Financial and Legal Priorities

Hajer argues that in order to assess a discourse we need to look not just at stated policies but also at investment strategies.[94] Treasury assumes a critical role in the assessment of institutional priorities and the distinction between a policy issue and a resources issue fades. The budgetary process offers a lens through which to distinguish underlying values—political, economic and social—insofar as they reflect in specific decisions to raise revenue and distribute expenditure.

This can be explained by evaluating the financial priorities of institutions and criteria for delivering transport solutions in Pakistan. Studying budgetary documents will help to investigate how much money was allocated in different years for road building, public transport and non-motorized transport. Criteria for funding from international institutions will help to explore their values and methods for adopting and implementing different programmes.

Past policies and decisions may also become embedded in the structure of the law. Policies grounded in law are backed by the coercive power of the state and hence are more durable. They give a signal to all actors of what has to be done and what cannot, duly supported by a system of rewards and penalties.

In this context it will be useful to describe not only past policies but also ordinances and laws. The important thing in relation to path dependence is to explore where the rules came from, and how they were incorporated and implemented to form overall policy; it will be therefore useful to identify laws which support a particular path. These laws reflect the original thoughts of that time and elucidate the pattern of institutional evolution.

The research framework to fulfil the aims of the study and answer questions discussed earlier requires a research design and a series of research techniques. These are as follows:

Research Methods

The question of what methodology is appropriate to an analysis is fundamental in setting a course of action for the whole investigation. Theoretical principles drawn from the concept of 'path dependence' will be used as a framework to provide guidelines on how to approach the study. In order to apply these guidelines, the case study method has been selected that will focus on the following areas:

1. A review of policy documentation to assess transport policy in the period British India–2005 in order to determine if there was a 'critical juncture'.
2. An investigation to determine if a 'self-reinforcement mechanism' exists in transport development. Relevant documents were vital in exploring institutions, techniques and discourse fields in the transport policies.
3. An empirical investigation of 'the interdependent web of an institutional matrix'. This will explore the transport decision-making process to assess institutions and explain the nature of the power structure in different organizations with reference to a particular case study.
4. Semi-structured interviewing of key figures involved in transport policy at the national level and of key players in a case study of a local transport project so as to set out the field of discourse.

'Critical juncture and 'initial selection' of the path can be explained through historical analysis. The analysis will identify how, when and why international institutions became involved in the transport sector of Pakistan. Contextual analysis will help to explore whether past policies show path dependent tendencies. This historical analysis will reveal the 'initial technical analysis' which may have locked transport solutions into a possibly inferior path. This will help to explore how decisions were made in the initial period of Pakistan's existence. For example, if road-based transport policies were selected because of supremacy of economic activities, then the study needs to explore the economic justification, and to what extent that was achieved.

To explore 'institutions', 'techniques' and 'discourses', a case study method is selected to allow the investigation to retain the holistic and meaningful characteristics of real life events.[95] Like other research strategies, case study is a way of investigating an empirical topic by following a set of pre-specified procedures.[96] Case study investigators can focus on data not available to the statistician. That is, the investigators can concentrate on an intensive, detailed and free-flowing collection of data on interactions or traits that are perceived to be relevant and important to the study.[97] In order to carry out the investigation the following case study has been selected.

Lahore in general and the proposed Ring Road in particular are selected as a case study for in-depth investigation of the research question. The Ring Road is partially built and comprises some sections upgraded from previously existing roads. The Ring Road is proposed to be a six-lane divided facility partly flanked by service roads, with eleven interchanges on major roads and six bridges over railway and canal crossings. This road is proposed to be a limited-access, urban toll highway. The road will interconnect radial roads converging on Lahore and will play an important role in guiding the future growth of the city.[98] The government has claimed that this road would assist the future growth of Lahore and improve the urban transport condition of the city; however, this claim needs to be examined.

The selected city and case study is representative of the issue to be investigated. Lahore, the provincial capital of the Punjab, is not only the second largest city of Pakistan but also an important centre of its political power. Although each major city of Pakistan has a particular character, Lahore has acquired a central position in the country's political landscape because the Punjab has 55.6 per cent of Pakistan's population by virtue of which it has acquired a primary role in national policy making. In equal measure, the Ring Road is the largest transport infrastructure project in Lahore, proposed with the technical and financial assistance of international institutions. A case study of the Lahore Ring Road will help to investigate how this project was generated and who funded its studies and construction at various times in history. Of particular importance will be determining the role of international institutions and their definition of 'sustainability' in the context of these projects. It may be noted that in selecting a case study, all major cities and their urban transport projects proposed in different studies were taken into account. Selection was made in accordance with certain criteria, such as the province's representation in national policy making, the scale of the project, involvement of international institutions and government perceptions of the urban transport system's economic, environmental and social sustainability. In addition to the above criteria, there were practical considerations that shaped the selection of this case study. Lahore and the Ring Road appeared appropriate because of the ready availability of information, the convenience of conducting interviews with local experts since the project was proposed by Lahore's Traffic Engineering and Planning Agency (TEPA) and the Lahore Development Authority (LDA), and finally the author's personal interest in Lahore.

The discourse field will be explored in two ways. Firstly, by identifying 'storylines' in planning and policy documents since these may have stabilized the existing policy path. Secondly, conducting semi-structured interviews, in particular of decision-makers involved in transport (local and international bureaucrats and professionals), in order to understand their perceptions and beliefs. Interviews are a direct source of information of intangibles like values, beliefs, attitudes and the knowledge system. Semi-structured and open-ended interviews are useful for obtaining information from selected respondents. Open-ended questions are flexible and allow the interviewer to probe the subject, get into greater depth and clear misunderstandings. Open-ended questions 'allow(s) the interviewer to make a truer assessment of what the respondents really believe'.[99] Questions were composed in a semi-structured way covering all dimensions of path dependence in order to find how decisions were made for transport related policies and the proposed project. The semi-structured interview questions provided insights into the perception of reality by respondents in a real world situation in a subjective rather than an objective manner.

Selecting respondents was an important step in the research process. The first question was who to interview and on what basis was the interviewee to be selected? It was useful to select respondents who were linked to transport related policy making either in their individual capacity or through their respective organizations, and also those experts, decision-makers and actors who were involved in the case study project. The next question was who to select for interview from amongst this large pool of presumably well-informed people? To this end, relevant individuals in international organizations and at different levels of government with likely greater knowledge of policy decisions in transport were selected. These can be categorized as:

- Professionals deployed in Pakistan by international organizations such as the World Bank and JICA. However, the citation of the World Bank expert in later chapters is based on information and presentation provided in local workshops on transport.
- Key experts of the federal government in the Planning Commission, Ministry for Environment and the Pakistan Environmental Protection Agency.
- Experts of the Punjab government working in the departments of planning and development (P&D) and communication and works (C&W). In addition, other stakeholders in Lahore were also consulted on transport related issues of the city.
- Experts in senior positions in the Traffic Engineering and Transport Planning Agency (TEPA) and the Lahore Development Authority (LDA) of the Lahore City District Government (LCDG).

- Academics in the field of transport, urban and environmental planning at the local universities.
- Researchers in government institutes such as the National Transport Research Centre (NTRC) and Pakistan Planning and Management Institute and in a non-government research institute such as the Institute of Policy Studies.

Moreover, the perspective of a corporate entity, i.e., the Indus Motor Company Ltd. (Toyota and Daihatsu) was obtained from presentations made by them in local workshops.

NOTES

1. North 1990; Hall and Taylor 1996; Pierson 2000b.
2. Pierson 2004: 1–2.
3. David 1985; Arthur 1988b; North 1990.
4. North 1990: 100; Berman 1998: 379.
5. Yin 1994: 32.
6. David 1985 and Arthur 1987.
7. Paul David 1985: 332–7.
8. Arthur 1994: 25.
9. Arthur 1988a: 10.
10. Arthur 1988b: 96.
11. Arthur 1988a: 10.
12. Douglas North 1990: 94.
13. Ibid., 103.
14. Ibid., 95.
15. Ibid., 98–99.
16. Pierson 2000a: 76.
17. Denzau and North 1994: 4.
18. The concept of 'mental models' goes back to antiquity but the phrase was coined by Scottish psychologist Kenneth Craik in the 1940s. He refers to the term as the semi-permanent tacit 'maps' of the world which people hold in their long-term memory, and the short-term perceptions which people build up as part of their everyday reasoning process (Senge et al. 1994: 237).
19. Denzau and North 1994: 22–26.
20. Ibid., 23.
21. Arthur 1988c.
22. Arthur 1994: 50.
23. Pierson's 2000a: 75.
24. David 2001: 16.
25. Arthur 1994: 50.
26. Arthur 1988b: 85.
27. Ibid., 96.
28. Peck 1996; 1999.
29. Woodlief 1998: 410.

30. Ibid., 411.
31. Pierson 2000a: 72.
32. Arthur 1988a: 26.
33. Pierson 2000b: 263.
34. David 1985; Arthur 1988a; Sewell 1996; Pierson 2000a.
35. David 1985: 332.
36. Pierson 2000a: 91.
37. Ibid., 73.
38. Mahoney 2000; 2001.
39. Woodlief 1998: 430.
40. David 1985: 334.
41. Jackson 1985.
42. Arthur 1994: 119.
43. North 1990: 95.
44. Mahoney 2000: 517.
45. March and Olsen 1989; Thelen 1999; Mahoney 2000.
46. Pierson 2000a: 77.
47. Scott 2001: 138.
48. Ibid., 81.
49. Hoberg 1991.
50. DiMaggio and Powell 1991b; Lodge 2003.
51. Gaventa 1980.
52. 1990.
53. North Dimitriou 1992.
54. Ibid.
55. North 1990
56. Daly 1996: 5–7.
57. Ibid.
58. Ibid., 6.
59. Meadows et al. 1992.
60. North 1990.
61. Hajer 1995.
62. Woodlief 1998: 430.
63. Pierson 2000a.
64. Arthur 1988b: 86.
65. Woodlief 1998: 410.
66. Ibid., 411.
67. March and Olsen 1984: 745.
68. Brain Arthur 1988a; 1994.
69. Pierson 1993: 599.
70. Ibid., 602.
71. Ibid.
72. Arthur 1994; North 1990.
73. Hajer 1995.
74. Thelen 1999 and Mahoney 2000.
75. March and Olsen 1989; Torfing 2001.
76. Wallace 1980: 13.
77. Pakistan ranked seventeenth from amongst 163 countries in term of corruption in 2005 according to Transparency International www.transparency.org. The ranking defines corruption as the abuse of public office for private gain, and measures the degree to

which corruption is perceived to exist among a country's public officials and politicians.

78. North 1990: 95.
79. Considine 1994: 97.
80. Govt. of Pakistan et al., 2001.
81. Hajer 1995: 63.
82. Gusfield 1981.
83. Dimitriou 1992: 167.
84. Arthur 1994 and North 1990.
85. Vigar 2002: 17.
86. Hajer 1995: 65.
87. Ibid., 44.
88. Hajer 1995.
89. Vigar 2002.
90. Ibid., 109.
91. Ibid., 158–9.
92. Ibid., 158.
93. Mazza and Rydin 1997.
94. Hajer 1995.
95. Yin 1994: 3.
96. Ibid., 15.
97. Hartman and Hedbolm 1979: 353.
98. LDA 1997: 6–40.
99. Cohen and Manion 1989: 313.

3

Pakistan's Transport Policies in the Historical Context

The historical overview of transport policies is undertaken by dividing the period under review into three time frames: pre-independence British India,[1] 1947–1990 and 1991–2005. Each of these periods provides a time frame of greater than ten years which is necessary to study and assess the dynamics of policy change.

Transport development in Pakistan since after independence is particularly important to document because the period represents and symbolizes the aspirations of the newly independent country.

TRANSPORT DEVELOPMENT IN BRITISH INDIA

The historical review of transport development in British India starts with the construction of the railway system. While debate in Great Britain on constructing railways in India started as early as 1840, it was in 1853 when the first goods train operated from Bombay to Thane, followed in 1854 by a passenger train from Howrah to Hoogly. The region that would later become part of Pakistan saw the introduction of railway in 1861 with the building of a track between Karachi and Kotri but up until 1865, important cities of this region were not connected by railway to the rest of the subcontinent. The railway network in present-day Pakistan was extended to the Afghanistan border and up to Zahidan (Iran) in 1878 and 1918, respectively.[2] In brief, British India had an extensive network of railways of 41,000 miles in 1944; of this, 8070 miles was in the area that now comprises Pakistan.[3] Locomotive workshops near Lahore, established in 1904, met the needs of the extensive railway network in the region.

Besides intercity railway development, rail transport also started to develop in urban areas. The first steam tramway started operations in Karachi in 1885 along with the horse-drawn tram but all these were eventually converted to petrol engine trams in 1908. Urban tramways and intercity railways were supported by buses providing feeder services.

In terms of the road network in British India, by the end of 1938 there were 19,000 miles of good quality municipal or urban roads along with 65,000 miles of regional or rural roads. Urban roads were managed by municipalities whereas intercity roads were the responsibility of the Public Works Departments (PWD) of each provincial government.[4] Vakil explains that the modern road network was limited to large cities and covered only a small area in comparison to the size and population of the country.[5] In summary, the Indian subcontinent under British rule had a well organized transport system, with an extensive rail network supported by a proportionately smaller urban road network. This discussion does raise the question of how a primarily rail-based transport system in British India impacted various economic and social activities.

The study found that quite apart from the strategic, commercial and economic benefits gained by the British in developing the railway in India, many economic and social advantages also accrued to the subcontinent. Macpherson narrates that about £95 million were invested by British companies in Indian guaranteed railways between 1845 and 1875.[6] According to Thorner,[7] large investments helped to establish an extensive network of railways which then played an active role in the development of both agricultural and industrial production. In the agricultural sector, the railways helped bring more land under cultivation, while for industry it represented the largest and most important modern enterprise in India, involving heavy machinery, precision techniques and advanced communication. In contrast, Shelvankar believed that the introduction of railways in India only helped to increase the supply of raw material intended to keep the wheels of British industry in motion.[8] Mitchell stated that in spite of the presence of rich natural resources, India's industry remained underdeveloped because the establishment of railways facilitated export of these resources to Great Britain as raw materials.[9] He found a huge disparity between potential resources and actual development which widened over time with the expansion of the rail network. Headrick argued that the impact of railways on the Indian economy was very different from that in Western countries because of the colonial context in which they were built.[10]

In terms of industrial development, Thorner found that railway development brought new consumer goods to the countryside that included matches, kerosene, lanterns and utensils.[11] Due to earnings from large exports of agricultural produce, people had the financial resources to purchase whatever they wanted. Similarly, mining and related industry also got a boost at that time. Modern cotton and jute spinning and weaving mills were established, and an engineering industry based on iron and steel was established in the early twentieth century. Food related industries such as vegetable oil and sugar mills were also set up. All these industrial activities were the consequence of railway development in large areas of the country. Thorner compared the impact of the railway network on industrial development amongst several countries.[12] He found that for leading industrial nations—particularly the USA, Russia, Germany, UK, France and even Canada and Australia—railways were the veritable dynamo for the industrial revolution. However, the situation was different in the case of India which in comparison to the above countries was relatively less industrialized. On a similar line, Headrick argued that 'tremendous decline in inland transport costs…and sharp increase of revenue from passengers did not contribute…to the industrialization of the subcontinent, but to its dependence on British industry and their engineers'.[13]

Vakil found that railway development accelerated India's internal trade which was fifteen times larger than its external trade.[14] It also supported movement of crops within the country and to the ports from where it was exported. However, Vakil's analysis shows that the commercial nature of the railway policy was mainly responsible for the rapid implementation of the industrial and trade policy for the country.[15]

Vakil found that the railway was one of the largest employment sectors in British India.[16] According to the Acworth Committee Report, the railways employed 717,000 persons in 1921; among them 710,000 were Indians and only 7000 were Europeans, a proportion of just 1 per cent.[17] However, Thorner stated that companies operating the railways in the subcontinent were not inclined to provide technical training to the Indians because that would allow them to hold higher staff positions.[18]

The development of railway also influenced the social system of British India. Thorner pointed out that even though the British made laws, and developed the administrative, irrigation, public health and educational systems in India,[19] it was railways which had the most significant and far reaching impact on British India. In his analysis of village communities, Thorner found that the railway had a more negative than constructive effect on Indian society.[20] Prior to the introduction of railways, village communities had close social and economic ties, but this structure was displaced by aggressive moneylenders, crop merchants and rent-receiving landlords.

Notwithstanding this debate, it would be absurd to deny the role of railways in the development of the Indian subcontinent. Thorner believes that independence in 1947 provided an opportunity to both India and Pakistan to formulate new railway policies which might have brought them much further along the path of industrialization.[21]

PAKISTAN'S TRANSPORT POLICY 1947–1990

A country's transport policy encapsulates its vision for development of transport, and should reflect the government's aspiration and priorities. However, a comprehensive statement of transport policy has been rare in Pakistan during the period 1947 to 1990. During this period Five-Year Development Plans were formulated which represented the federal government's priorities for development in different sectors, including transport. Policies on transport will therefore need to be inferred from the Five-Year Development Plans during the period 1947 to 1990.

DEVELOPMENT PLANNING IN PAKISTAN

Pakistan's history of development planning started before the partition of British India, at the time when the department of planning and development was first created soon after the Second World War. When Pakistan came into being in August 1947, its federating provinces already had a number of development projects that were prepared before independence;[22] soon thereafter, a department of development was set up under the federal ministry of finance.[23] Planning in Pakistan received an added boost in 1950 when Pakistan joined the Colombo Plan for Co-operative Economic Development in South and South-East Asia and remained its member for six years. Under the Colombo Plan, transport infrastructure such as road transport, civil aviation, shipping and port facilitates received the highest attention.[24] However, in real terms, Pakistan's development planning started with the preparation of the First Five-Year Plan in 1955. This exercise is still continuing with the development of the eighth Five-Year Plans and three Ten Year Perspective Development Plans. Currently the Medium Term Development Framework (2005–10) (a new name instead of the Five-Year Plan) is being implemented under the overall objectives of the Ten Year Perspective Development Plan (2001–11).

The federal Planning Commission, the main planning and policy making authority, prepares the Five-Year Plan and other mega projects approved by the National Economic Council (NEC) or the Executive Committee of the National Economic Council (ECNEC). The Commission is also responsible for preparing the federal annual development plan and for reviewing and prioritizing provincial development projects.[25] In addition, the Planning Commission controls official 'development assistance' from international and bilateral assistance agencies.[26] By virtue of the control it exercises over domestic funds, the country's foreign exchange reserves and the large quantum of foreign assistance, Pakistan's federal government has a dominant position in the three-tiered hierarchy of government.[27]

Transport Priorities in Pakistan from 1947 to 1965

The 1947 Pakistan was an underdeveloped area with its economy primarily dependent on agriculture. Most major ports in the subcontinent and its centres of trade and industry became part of India. However, the country did inherit a comprehensive network of well-developed rail and road network notwithstanding the fact that historically investment in areas comprising Pakistan had been low.[28] The railways were the most valuable capital asset of the country and its largest, single national commercial undertaking. At that time Pakistan Railway (North Western Railway) was the largest provider of transport for goods and passengers in Pakistan.[29]

The initial difficulties the country encountered soon after independence were largely surmounted within two years and economic conditions started to improve by 1949. To coordinate nation building schemes, a development board, a planning advisory board and an economic committee of the cabinet were set up in 1948. These organizations were replaced by an economic council and a planning commission in 1951.[30] Notwithstanding the ambitions for development, Hasan, a former head of the Planning Commission, described the apprehensions in accepting foreign aid and loans:

> Three times in 1950s, the Government [of Pakistan] declined the proposed American assistance. The cabinet also split in the same year over the question of going to the World Bank for loan[31]...The Commonwealth aid, promised through the Colombo Plan was, however, accepted unhesitatingly[32]...Even, Ghulam Mohammad [Governor General of Pakistan 1951–55], the first proponent of foreign aid [as a Finance Minister] was opposed to foreign interference in development planning, which in time became a corollary of foreign aid.[33]

A Six Year Development Plan was prepared and incorporated into the Colombo Plan in mid-1951. A review of the transport section of the plan shows that civil aviation, shipping, port facilities and road transport were developed from negligible beginnings.[34] However, the plan retained railway as the backbone of passenger and freight transport as was evident from the inclusion in the plan of a 'Railway Rehabilitation' programme and the planned construction of 100 miles of new railway lines.

Waterston points out how the Six Year Plan was prepared within a period of only three months.[35] The government acknowledged that the plan was prepared in the absence of essential information and basic statistics which resulted in an improper assessment of national resources.[36] The reason behind this shortcoming was a serious shortage of technical people and lack of adequate capacities of federal, provincial and local government organizations of the newly independent country.[37] Arrangements were therefore made under a grant from the Ford Foundation to enlist foreign consultants who were to be supervised in their planning work for Pakistan by experts from the Harvard University.

The Harvard Group had three main tasks:

> ...(the first task was) to help the Planning Board to prepare the First Five-Year Development Plan; the second, to organize on job training for Pakistani staff in a central planning body which would enable them to carry out research and planning activities without outside assistance; the third was to provide advice to the government on current economic, financial and fiscal policy.[38]

How and up to what extent the Harvard Group achieved the second and third objectives will be discussed in the next chapter but for the present we need to describe the first objective of the Harvard Group, and examine how that objective affected the pattern of transport development in Pakistan.

The main objective of the First Five-Year Plan (1955–60) prepared in 1956 with the help of the Harvard Group was:

> to develop the resources of the country as rapidly as possible so as to promote the welfare of the people, provide adequate living standards and social services, secure social justice and equality of opportunity and aim at the widest and most equitable distribution of income and property.[39]

Accordingly, a transport plan was formulated and allocated 17.8 per cent[40] of the total resources. The plan provided sizable expenditures for the railways and for expansion of the road network. The First Five-Year Plan accepted that:

> ...the backbone of [West] Pakistan's transport system is a broad-gauge railway network. It is a system of main lines, one in each of five parallel river valleys, interlinked and stretching from the coast to Afghanistan and India's frontiers.[41]

However, the Plan proposed that 'in [West] Pakistan a powerful railway system and growing road transport system operate side by side and should complement each other'.[42] Keeping in mind this policy, 512 million rupees (70 per cent of total land transport expenditure) were allocated for Pakistan Railway (North Western Railway) and 220 million rupees (30 per cent of total land transport expenditure) for road transport during the plan period of 1955–60.[43]

A road transport board financed by federal and provincial governments was set up to achieve rail-road coordination to achieve a ratio of 25 and 75 per cent in favour of the railway. It is important to note that although resources were allocated in favour of the railway, the Plan proposed that the ratio of road to railway would increase to 25:75 compared to 10:90 in 1947. Moreover this Plan did not propose any extension of the railway network. On the other hand 1800 miles of new roads were planned to be constructed along with the improvement of 2000 miles of existing roads.[44] The Plan established a road building research laboratory at the provincial level but research was limited in scope, only to investigate new specifications for road construction using locally available material. The Plan also encouraged importing road building machinery.

The First Five-Year Plan promoted road transport by establishing the (West) Pakistan Road Transport Board in 1957. For promoting road-based public transport, the Karachi Road Transport Corporation (KRTC) was established to operate bus-based urban public transport in Karachi. The Harvard Group had also worked with the provincial governments to formulate and execute their respective plans in light of the First Five-Year Plan. Waterston described how in the absence of qualified Pakistani staff at the provincial level, foreign consultants acted as advisors to provincial officials when they interacted with the national economic council, the national planning board and the ministry of finance.[45] He stated that a substantial amount of technical assistance was furnished to the provincial planning departments by the United Nations, the UN Food and Agriculture Organization, the Colombo Plan Organization, the US government and resident representative of the World Bank. These advisors were involved in the preparation and execution of individual development projects as well.

The First Plan was executed mainly from the country's own financial resources. However, 23 per cent of allocated resources were obtained through foreign aid and loans, most of it bilateral and much of it from the United States. This was in addition to US technical, food and military assistance which started in 1952. But multilateral aid has been substantial ever since 1952 when the World Bank started to operate the earliest and largest programmes for infrastructure investments in Pakistan.[46] The World Bank aid also gathered momentum during the second half of the 1950s totalling US$800 million as against US$126 million in the first half of the 1950s.[47] Brecher and Abbas argued that by the beginning of First Five-Year Plan, Pakistan's reliance on foreign assistance had become virtually inevitable.[48] They found that during the early period of the development plan, the industrial sector was the single largest recipient of foreign project aid, closely followed by transport and communications.[49]

In brief, the First Five-Year Plan was balanced for both railways and roads. The Harvard Group played a key role in establishing the foundation of development in Pakistan, including transport, at both the federal and provincial levels of government. However, it opened the doors to reliance on foreign technical and financial assistance to formulate development policies in Pakistan.

The main objective of the Second Five-Year Plan (1960–65) was re-stated in light of the objectives of the First Five-Year Plan:

> The nation aspires to a standard of living for all its people as high as can be achieved with the resources available to it; equitable distribution of wealth; education of all in accordance with their talents; victory over disease; adequate facilities for transport and communications so that the nation may be effectively unified economically and socially; and evolution of the national culture in literature, art and science.[50]

To attain the stated objective, the Second Five-Year Plan developed a long-term transport development plan to postulate a sound pattern for the country's transport system.[51] The Plan allocated 17.6 per cent of total plan development expenditures to transport, almost the same amount allocated during the First Five-Year Plan. In the allocated resources for land-based transport, the railway accounted for 44.7 per cent as compared to 55.3 per cent allocated for roads. Hence, the Plan became the first planning document in Pakistan in which the roads sector was given priority over railways. The considerations for this prioritization were stated in the Plan as:

> Road transport is particularly suited to the conditions and requirements of Pakistan[52]...the motor vehicle is more adaptable than the railways to varying degrees of traffic intensity and permits a greater degree of speed and efficiency in haulage over short distance[53]...there is close relationship between the volume of transport and the level of economic activity because each depends upon the other.[54]

With these arguments in mind, which help to create a widespread view amongst decision-makers in favour of road development, the Second Five-Year Plan initiated a new era of road construction in Pakistan.

Major cities of Pakistan were also inclined towards construction of new roads in the belief that it would accelerate economic development. In parallel to road construction, road-based transport was also promoted. The Second Five-Year Plan allocated Rs75 million to the [West] Pakistan Road Transport Board to introduce 500 new buses in its fleet for intercity public transport.[55] For intra-city or urban transport, Rs32 million was allocated to the KRTC for building up a fleet of 1200 buses, procuring 700 vehicles in addition to the 500 obtained in the First Plan period.[56] The Second Plan encouraged the private sector to come forward and run public road transport.

The Second Five-Year Plan supported the inclusion of the Karachi Circular Railway (KCR) as the first (and last to date) urban rail public transport project in Pakistan. KCR was planned to serve the whole of Karachi including the city's periphery. It was projected as a regular, cheap and efficient transport for the residents of Karachi. The plan allocated Rs23 million for this project.[57]

The Second Five-Year Plan depended heavily on foreign assistance and investment.[58] The Plan expected the government to generate 42 per cent of total expenditure from foreign assistance in the form of aid, loans (21 per cent) and foreign investment (21 per cent), in addition to defence and humanitarian aid.[59] The United States provided the bulk of this aid (78.7 per cent) followed by the World Bank (7.7 per cent).[60]

Waterston explained that in contrast to its central role in drafting the First Five-Year Plan, the Harvard Advisory Group made lesser direct inputs in the Second Plan.[61] However, he stressed that the character of the Second Five-Year Plan was materially influenced by the Harvard Group through advice, background papers, staff comments, and by actual drafting or redrafting of sections of the Plan. It is fairly evident therefore that the Harvard Advisory Group played a central role also in the preparation of the Second Five-Year Plan.

In summary, the Second Five-Year Plan gave a new direction to transport development in favour of roads. Although the basis for this direction was laid down in the First Five-Year Plan (1955–60), the process was accelerated in the Second Plan. This argument is based on the judgement of Waterston, who claimed that the Second Five-Year Plan merely extended the course of development along lines laid out in the First Plan.[62] He also explained that the current preference of donor countries and international lending institutions for comprehensive development plans to be the basis for their aid also helped strengthen the ongoing trend of planning practices in Pakistan.[63] Arguably the choices made in the First and Second Five-Year Plans were significant because they conformed to the preferences of international development institutions. Most advisors of the Harvard Group came from the USA and Europe, and even though their level of competence was high, none had prior planning experience in underdeveloped countries.[64] Also, planning practices were limited to a project-by-project approach as opposed to comprehensive and aggregative planning encompassing the entire economy.[65]

Overview of National Transport Priorities in Development Plans

Various Five-Year Plans were reviewed in order to determine the principal paradigm of transport development in Pakistan. The discussion will run mainly on three components: the percentage of the overall development budget allocated for the transport sector, the money allocated for transport infrastructure and the impact of these policies on transport infrastructure.

Figure 3.1 shows the percentage of total money spent on the transport and communication sector in the sequence of Five-Year Development Plans. In the First Five-Year Plan (1955–60), 22.2 per cent of the total development budget was allocated for transport and communication which rose to 23 per cent in the Fifth Five-Year Plan (1978–1983). Subsequently the proportion remained between 17 and 18 per cent of the total development budget until the Eighth Five-Year Plan (1993–98). In summary, one-fifth of Pakistan's development budget was allocated to the transport and communication sector over the last fifty years. The graph shows that governments throughout recent history have made serious

FIGURE 3.1

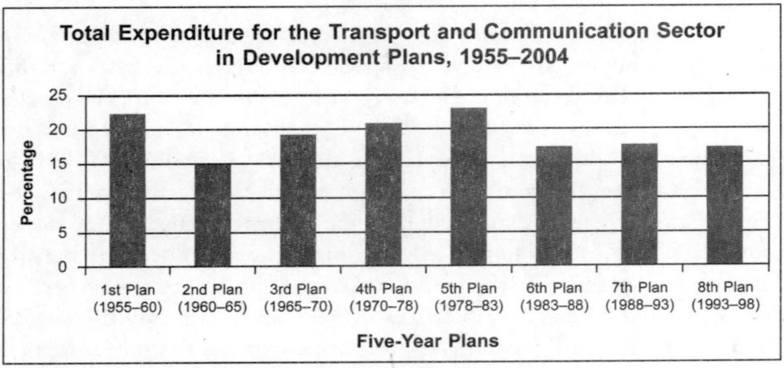

Source: Govt. of Pakistan, Federal Bureau of Statistic 1997a: 341

efforts to develop the transport network in Pakistan. However, the line chart (Figure 3.2) demonstrates the mode of transport which received government priority in various Five-Year Plans.

Figure 3.2 compares the money allocated to land-based transport from within the overall transport sector budget from 1955 to 2001.[66] It can be seen that the railway was accorded priority only in the First Five-Year Plan (1955–60) when it was allocated 65.3 per cent (Rs683 million) of the total development budget as compared to 34.7 per cent (Rs360 million)

FIGURE 3.2

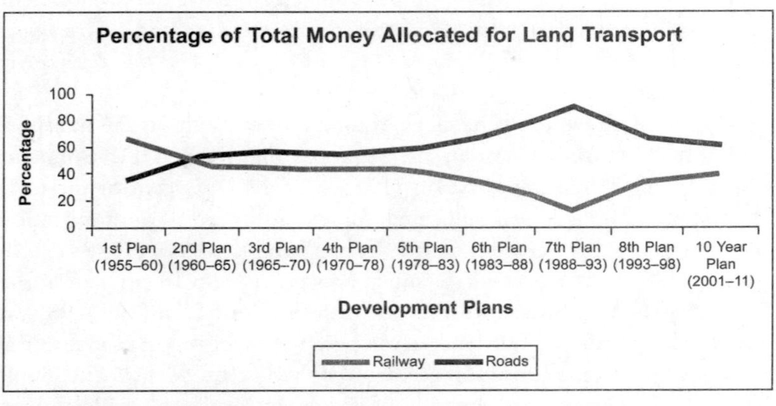

Source: Data extracted from all Five-Year Plans

for road infrastructure. The percentage of money allocated to railways gradually decreased and reached a low of 41.7 per cent (Rs6773 million) in the Fifth Five-Year Plan (1978–83). From then on it continued to decrease more sharply, and was down to 12.5 per cent (Rs9000 million) by the Seventh Five-Year Plan (1988–93). However, over the last two development plans the allocation for railway development has gradually improved. In contrast, the percentage of money allocated for road development increased from 34.7 per cent (Rs360 million) in the First Five-Year Plan (1955–60), and peaked in the Seventh Five-Year Plan (1988–93) when it reached 87.6 per cent (Rs29,592 million). After that, this figure started to decline and has dropped to 60.7 per cent (Rs163 billion) in the Ten Year Perspective Plan for 2001–2011.

The review of transport development as reflected in successive Five-Years Plans clearly indicates that resource allocations were in favour of road development. The First Five-Year Plan (1955–60) was an exceptional case when the comprehensive network of the railway system inherited from the British was declared the backbone of transport.[67] The result of these policies can be seen in Figure 3.3. The chart shows that the road length in the country increased by more than four times over what it was in 1947, in contrast to the railway which remained stagnant at approximately the same length in kilometres as in 1947. This road development also resulted in a sharp increase of motorization; 3,838,167 vehicles in 1995–96 as compared to 21,209 vehicles in 1947–48.[68]

FIGURE 3.3

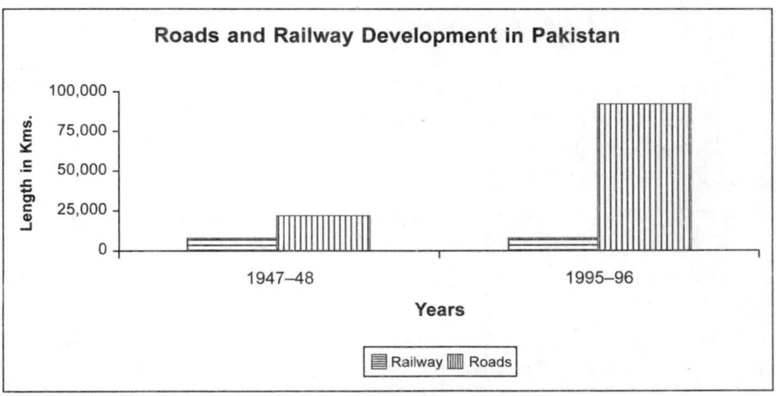

Source: Govt. of Pakistan, Federal Bureau of Statistic 1997a: 4-30

It was acknowledged from the outset that there was an acute shortage of appropriate expertise, reliable data and supporting laws to formulate and implement transport policies in Pakistan. Yet, in spite of the massive spending on the road and railway systems, less than 1 per cent of allocated money was spent on research and institutional capacity building (for instance, building organizations and skills) throughout the history of Pakistan.

PATH DEPENDENCE: EXPLORING THE CRITICAL JUNCTURE

Thorner believes that a railway policy different to that adopted under colonial rule and more specific to the needs of a newly independent Pakistan could have been designed in 1947 which might have helped in the country's rapid industrialization.[69] It remains to be seen why Pakistan failed to adopt such a policy. Within the theoretical framework, research finds a 'critical juncture': a period in which certain social forces combine to change the pre-established pattern of transport in the country and shape the future path of transport policy. The combination of these social forces came about at a critical juncture to push the country into adopting road construction in preference to the railway as the main policy instrument for intercity transport.

When Did the Critical Juncture Occur?

A review of transport priorities and policies shows that the period over which transport policies changed in favour of road development was between the mid 1950s and early 1960s when new transport priorities were laid down. Figure 3.2 confirms this as it shows that budget allocation for road development was given priority in the Second Five-Year Plan (1960–65) in spite of the fact that an extensive railway network was already fulfilling the transport needs of Pakistan. We argue that the time period 1955 to 1965, during the First Five-Year Plan and Second Five-Year Plan was a critical juncture in the transport development of Pakistan.

What then were these social forces and how did they shape the future path of the country's transport policy? It was argued that during the First and Second Five-Years Plan, international development institutions such as the Harvard Advisory Group and foreign consultants were the two major social forces that shaped transport programmes to achieve rapid economic development. Details of their intervention and how they shaped the future path of Pakistan's transport policy will be discussed later.

The history of transport in Pakistan provides evidence of two components of the 'critical juncture' concept described by Mahoney.[70] First, there must be choice when a particular option is adopted from amongst two or more alternatives. In the beginning, Pakistan had a choice to opt for either railway or road development to achieve a high level of mobility. It may be recalled that prior to the commencement of development plans in 1955, Pakistan had well established assets of a railway system that spanned the whole country, while its urban areas relied mostly on non-motorized and bus-based public transport. However, a historical review of transport priorities in Pakistan reveals how competition amongst several modes of transport started during the 'critical juncture' time period, resulting in the dominance of road development.

Transport priorities established during the early phase of development planning defined the general direction of transport development for the period that followed. To begin with, Pakistan had the option to either continue with emphasis on railways or promote road development; the government opted for the latter and a small initial investment in roads satisfied the needs of the people. However, because foreign consultants involved in the First and Second Five-Year Plans (1955–65) favoured road development, road transport has now come to be the main transport infrastructure of the country (see recent Medium Term Development Plan for 2005–10 discussion in Chapter 4). The same initial road development concerns prompted the government to prioritize road-based transport in the country. This initial bias of the experts was enough to push virtually the entire newly independent developing country into road development. This occurred in spite of the fact that more sustainable alternatives such as the railway and a large number of non-motorized trips still exist in Lahore (60 per cent)[71] and Karachi (50 per cent).[72] In contrast, Hong Kong continued to invest in the existing railway network for use as an urban rail system (in the case of Kowloon Canton Railway).

The second component in Mahoney's description of 'critical juncture' is that once a particular option is selected, it becomes progressively more difficult to return to the initial point when multiple alternatives were available.[73] This component of 'critical juncture' fits the situation of inter-urban and intra-urban transport development of Pakistan and will be discussed in detail later.

At this stage it is important to consider if there was only one critical juncture in establishing new transport priorities. What about the Seventh

Five-Year Plan (1988–93) when railways were again allocated more money as compared to the previous plans? (See Figure 3.2) It was shown that there was only one critical juncture during 1955–65 when road development started to receive priority over other modes of transport. Although Figure 3.2 shows that allocation of money for the railways increased in the overall budget for transport development during the Seventh, Eighth and Tenth Five-Year Plans, this increase was prompted by the need to save the railway system from being totally abandoned and was not intended to bring the system into competition with road development (see recent Medium Term Development Plan 2005–10 discussion in Chapter 4).

Why the Critical Juncture Occurred? Explaining the Role of International Development Institutions in Transport Development in Pakistan

A significant feature of Pakistan's central planning in preparing the First Five-Year Plan (1955–60) was extensive use of foreign technical assistance. Soon after the establishment of the Planning Board in 1948, it was recognized that there was a dearth of qualified staff to formulate development plans.[74] The ministry of economic affairs therefore asked the Ford Foundation to acquire for Pakistan the services of economists who were to be recruited and supervised by the Graduate School of Public Administration of Harvard University.[75] A cosmopolitan and versatile group of twelve full-time advisors and five consultants started work on the Five-Year Plan in 1954.[76] Top experts were engaged with considerations of high emoluments and other expenditures being set aside.[77] Dean Mason, who headed the Harvard Group, was an internationally renowned economist and an advisor to the American administration and the World Bank. David Bell, who worked for four years in Pakistan, went on to become the director of budget in President Kennedy's administration and was later the Director of USAID. The Harvard Group also included prominent professors like Prof. J.K. Galbraith from the Harvard University, economists such as John C. Eddison and researchers such as John H. Power, Mark W. Lelserson, Philips S. Thomas, Keith B. Griffin and Stephen R. Lewis,[78] all of whom helped in preparing Pakistan's development plans.[79] The Harvard Group was headed by a chief advisor, and each member was assigned a specific section. However, they were all considered employees of the Planning Board and acted as heads of their respective sections.[80]

The task of the Harvard Group was to help the Planning Board in preparing a Five-Year Development Plan for the country and to develop the institutional capacity of the planning organization to carry out research and other activities without external assistance. The Harvard Group remained an integral part of the Planning Commission for nearly a decade and was intimately involved in preparing the First and the Second Five-Year Plans. When the contract of the Harvard Group ended in 1965, which is when the World Bank took over, the Planning Commission had benefited from the services of a total thirty-five foreign advisors and twenty consultants.[81]

Noman also found that the bilateral influence of the 1950s shifted to a more multilateral arrangement in the early 1960s when institutions such as the World Bank started to hold sway over Pakistan's development plans.[82] He observes that the contribution of the aid and loans received by Pakistan in the 1950s was more in the nature of policy formulation. Hasan, a deputy chairman of the Planning Commission in the 1960s, explained that the Harvard Group was helpful in securing guaranteed loans for Pakistan from the World Bank.[83] He argued that the Harvard Group in collaboration with the World Bank shaped early development programmes in Pakistan.[84] The involvement of international development institutions and their development paradigm was projected as a success of development in Pakistan such as:

> Pakistan may be on its way toward an economic milestone that so far has been reached by only one other populous country, the United States.[85]
>
> The survival and development of Pakistan is one of the most remarkable examples of state and nation building in the post-war world.[86]

The role of the Harvard Group was so significant that the path they set for Pakistan in the 1960s is followed even today,[87] this being especially true for transport. However, an example from the irrigation sector of Pakistan would be helpful in understanding the involvement of international development institutions and their foreign consultants during First and Second Five-Year Plans period.

In Pakistan (in contrast with India), participation of Pakistanis in the huge Indus Basin works (in early 1960s) was nominal—there were tiers of foreign (American) consultants, engineering plans were made in America and the contractor (an American) was selected by the World Bank who also presumably kept the accounts. There were almost no Pakistani engineers, contractors or manufacturing firms involved in the project, save a chief engineer who was essentially there for public relations. The cost was excessive.[88]

It is clear from Figure 3.4 and research conducted by Hasan that the heavy dependence on external technical and financial resources emerged early in Pakistan's history and has shown no declining trend since.[89]

External resources are important for developing countries in the process of their development. In the case of Pakistan, the signing in 1954 of the Mutual Defence Agreement with the US opened the way for large scale military and economic assistance. During the first three years of the First Five-Year Plan, economic assistance totalled 2.8 per cent of the GDP of Pakistan.[90] A comparison of aid and loans during 1949 to 1959 shows that Pakistan received US$14.4 per capita foreign aid as compared to US$6.5 received by India.[91] In the Second Five-Year Plan, it was assumed that more than 40 per cent of public sector development funds would be arranged from external sources. In the Third Five-Year Plan (1965–70) following the 1965 war with India, foreign economic assistance tended to decline.

However, as Figure 3.4 shows, economic assistance again increased during the Fourth Five-Year Plan/Non-Plan Period (1970–78).[92] This time the major proportion of assistance came from oil exporting countries of the Middle East. Loans and grants of over US$337 million in the First Five-Year Plan (1955–60) increased to US$10.11 billion at the end of the Eighth Five-Year Plan (1993–98) (see Figure 3.4).

Husain argues that foreign borrowing and aid currently finance 80 to 90 per cent of development expenditure in Pakistan, and the nation's reliance on external flows of finance for development has increased steadily over time.[93] According to the Ten Year Perspective Development Plan (2001–11):

> Over the last many years Pakistan's major problem has been its excessive dependence on foreign capital for financing its development programmes. This has led to the accumulation of foreign debt which has now become a burden on the economy. Total external debt stood at $32.7 billion on 30th June, 2000 constituting 53.4 per cent of GDP.[94]

In summary, Pakistan received US$58 billion of foreign development assistance (in 1995 dollars, between 1960 and 1998), twenty-two adjustment loans from the IMF and the World Bank, a lucrative Cold War alliance with the US and multiple government development programmes. Pakistan was the third largest recipient of official

FIGURE 3.4

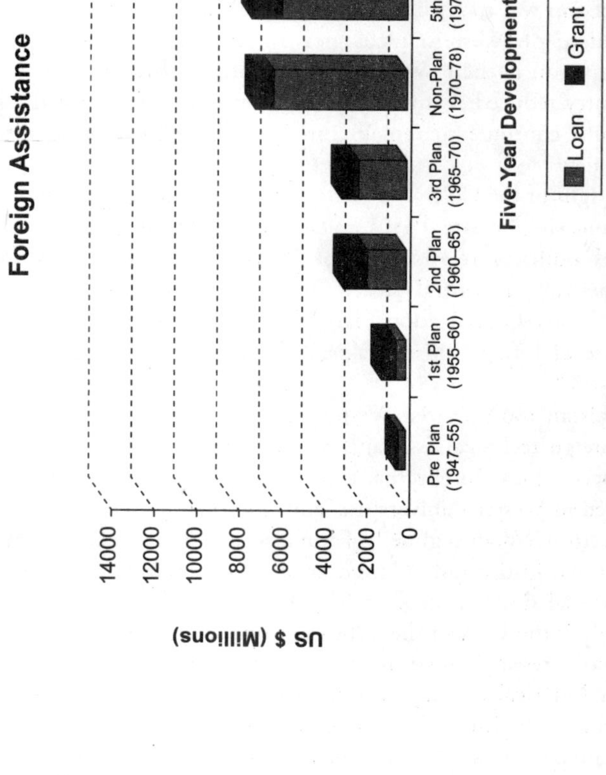

Source: Figure extracted from the data of all Five-Year Plans

development assistance in the world over the period 1960–1998 (India and Egypt are respectively first and second). The World Bank alone provided US$20 billion in loans from 1952 to 1999.[95] However, Easterly found that, whilst foreign aid and government programmes may have contributed to overall economic growth, they were an egregious failure at promoting social and institutional development.[96]

Some radical critics argue that this assistance was deliberately designed to prevent Pakistan from developing an independent policy capacity and hence keep it dependent on foreign financial and technical assistance. A development economist and academic Dr Mahnaz Fatima in a critical review says: 'A challenge: Pakistan's external dependence peaks at golden jubilee' argued that the USA had developed a Cold War strategy for the developing world to achieve two aims.'[97] The first aim was to ensure a country's financial dependence on external development assistance, especially on the Bretton Woods institutions, e.g. the World Bank. The second aim was to ensure that developing countries remain intellectually dependent on Western ideologies by offering these countries generous participation in their (Western) educational programmes. That is, even if a country reduced its financial dependence on international institutions, it would continue to remain intellectually subservient to the Western worldview.

In light of the Cold War strategy, the large amount of foreign financial assistance in the late 1950s and early 1960s was made possible by the special political relationship of Pakistan with the USA [due to its membership of Central Treaty Organization (CENTO) and South East Asian Treaty Organization (SEATO)].[98] This relationship resulted in an increase of foreign aid to Pakistan by up to 7 per cent of its GDP by 1964–65.[99]

Pakistan too acquired Western development ideas in consequence of the foreign technical assistance it received. This intellectual dependence produced a local 'intellectual elite' resulting from the education Pakistanis received in Western universities, and who then touted policies dictated by the Bretton Woods regime.[100] Fatima argued that intellectual subservience to external institutions allowed greater foreign domination in all spheres of national decision-making in Pakistan.[101] The local 'intellectual elite' created by the USA in the 1960s not only controlled major public sector economic research institutes in Pakistan and the research funds from North America, but also directed the research agenda and expression of thoughts at high-profile national conferences. This process blocked the alternative view in all development-related sectors.[102] Like Fatima, Huda also points out that foreign technical assistance in the early 1960s was mismatched to local needs and circumstances.[103] Further, Hasan found that it was doubtful if the advice of the World Bank given to Pakistan in its earlier years of development planning was good for the well-being of the country.[104] However, alongside an expansion in the World Bank's activities, its influence increased and 'its casual views came to be cited as the holy writ'.[105] He concluded that regardless of how good or otherwise the role of the World Bank had been, the fact is that it occupied a principal position in advancing a specific development paradigm in the country. Whether or not I agree with Fatima and Hasan's analysis 'across the board', the transport sector seems to confirm much of their findings. Along similar lines, Szyliowicz describes the history of transportation in Turkey, explaining how highway development assumed dominance over the country's railway network.[106] He referred to the famous saying of the Turkish President that, 'railways are for communists, highways for capitalists'.[107] This statement conforms to the data displayed in Figure 3.5. The graph demonstrates that financial assistance from the World Bank increased steadily in favour of road development from the mid-1950s to mid-1960s, which is the period during which Pakistan started getting technical and financial assistance from the international development institutions. Hasan explained that the main aim of the First and Second Five-Year Plans was to attract foreign aid[108] which is why foreign consultants matched the plan to the international funding that was available, as seen in Figure 3.5.

It is important to note that a large proportion of foreign financial and technical assistance was reserved for project-oriented development such as the construction of transport infrastructure rather than for building local policy-making capacity. Figure 3.6 validates this conclusion by showing that of the total aid received by Pakistan for all its Five-Year Plans, project aid was larger than that given as non-project aid.[109]

Ercelan has argued that there are three serious problems with this aid to projects.[110] They are: inappropriate and ill-conceived projects, direct adverse impacts on people's lives and livelihoods, and a growing and excessive fiscal burden on the national economy. In his research on Japan's Overseas Development Assistance (ODA) to Pakistan, Ercelan stresses that whilst foreign technical experts mostly identify development needs correctly, they usually respond with actions and priorities that are technically inappropriate and unsuitable for local conditions.[111] He also stressed that fifty years of Japan's ODA has placed a heavy emphasis on hard infrastructure rather than institutional development. In contrast with

FIGURE 3.5

IBRD and IDA Commitments to the Transport Sector (1950–65)

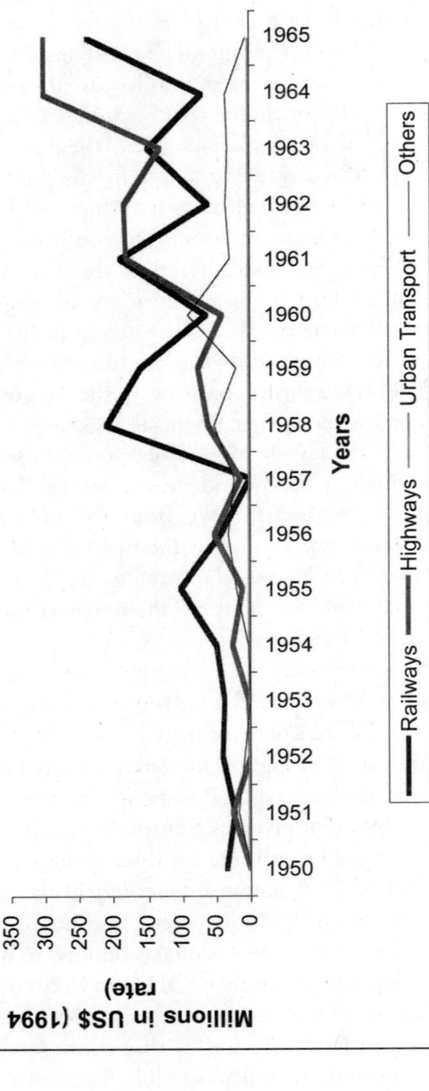

Source: World Bank 1994: Table A.3, 149

FIGURE 3.6

Aid for Projects and Non-projects

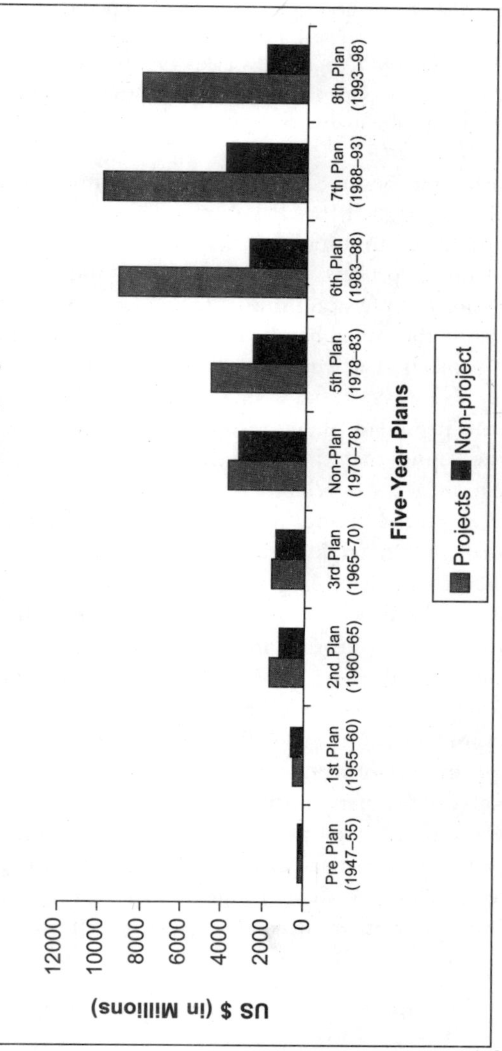

Source: Govt. of Pakistan, Federal Bureau of Statistic 1997b: 667–677

Ercelan's analysis, Hasan found that the Harvard Group failed to correctly identify the problems in the First Five-Year Plan.[112] He discovered that:

> The [First] Plan...was beautifully composed, superbly symmetrical in design and phrased in terse American functional language, but it lacked a soul. It was bereft of a theology recognizably Pakistani. It was evidently the work of craftsmen engaged to do a specific job. It did not relate to the milieu in which people lived and laboured or to their mood in 1955.[113]

In a famous article 'Seven Sins of Economic Planners' [during the 1950s and 1960s], Haq argued that Pakistan needed aid to enable it to utilize existing capacity of [transport] infrastructure.[114] However, he observed that the planning process was isolated from the society at large and existing resources of physical infrastructure [e.g. railway] were not utilized properly.[115] On the other hand, international development institutions focused on projects that would result in a high rate of growth. Qureshi argued that there was nothing wrong in the use of borrowed resources for development provided these resources were effectively invested.[116] However, he found that there was no organized programme designed specifically to meet the development needs of the country.[117] In short, the fruits of development did not reach the masses in Pakistan.

In general and in an historical context, international development institutions and associated foreign consultants played an important role in formulating the development plans of Pakistan. More specifically, a strong case can be made that involvement of foreign consultants and advisors and availability of large financial aid in the First and Second Five-Year Plans established a 'critical juncture'; this was a turning point in the transport development of Pakistan because it narrowed the range of possible transport outcomes in the country. A basic outcome of the international development institutions intervention was to completely replace the transport investment scenario, as can be seen in Figures 3.2 and 3.3. During this 'critical juncture' a specific transport option (roads and private vehicles, and their complementary institutions) was selected to take priority over other alternatives (such as railways).

NOTES

1. For most part of the seventeenth century and all of the eighteenth century, India was largely under the control of the East India Company, which first started operating in the subcontinent in 1600. However, British India as a single entity did not form until the second half of the eighteenth century. Some writers argue that British rule in India started in 1757 when they established their rule in Calcutta while others favour 1773, the year when the British parliament passed the Regulation Act and appointed Warren Hastings (1773–85) as the first governor-general; and yet others mark 1858 as the year British India came into being, when the East India Company was dissolved and responsibility for administration of India was directly assumed by the Crown and discharged through the India Office established under the British Department of State. This arrangement continued right up to 1947 when India and Pakistan gained independence.
2. http://irfca.org/faq/faq-hist.html#first (Indian Railway Fan Club Website).
3. Vakil 1944: 187.
4. Ibid., 190–192.
5. Ibid.
6. Macpherson 1955: 177.
7. Thorner 1951: 399–400.
8. Shelvankar 1940.
9. Mitchell 1942: 5.
10. Headrick 1981.
11. Thorner 1955: 213.
12. Thorner 1955: 201.
13. Headrick 1981: 188–189.
14. Vakil 1944: 187.
15. Ibid., 189.
16. Ibid., 189–190.
17. Cited by Thorner 1951: 398.
18. Thorner 1955.
19. Thorner 1951: 399.
20. Thorner 1951.
21. Thorner 1955.
22. Waterston 1963: 13.
23. Hasan 1999: 142.
24. Ibid., 151.
25. LaPorte 1992: 108.
26. Ibid.
27. Waterston 1963: 137.
28. Hasan 1998: 97.
29. Govt. of Pakistan, National Planning Board 1957: 487.
30. Ibid., 8.
31. Hasan 1999: 91.
32. Ibid.
33. Ibid., 93.
34. Govt. of Pakistan, National Planning Board 1957: 10.
35. Waterston 1963: 15.
36. Govt. of Pakistan, National Planning Board 1957: 9.
37. Ibid., 12; Waterston 1963: 22.

38. Waterston 1963: 34.
39. Govt. of Pakistan, National Planning Board 1957: 1.
40. Ibid., 15.
41. Ibid., 485.
42. Ibid.
43. Ibid., figure extracted from Table 7, 510.
44. Ibid., 46.
45. Waterston 1963: 38.
46. Brecher and Abbas 1972: 22, 66.
47. Hasan 1998: 119.
48. Brecher and Abbas 1972: 23.
49. Ibid., 25.
50. Govt. of Pakistan, Planning Commission 1960: 1.
51. Ibid., 278.
52. Ibid., 303.
53. Ibid., 304.
54. Govt. of Pakistan, Planning Commission 1960: 277.
55. Ibid., 305.
56. Ibid., 306.
57. Ibid., 287.
58. Development aid included private foreign investment as part of total foreign assistance required by developing countries; Brecher and Abbas 1972: 1999.
59. Govt. of Pakistan, Planning Commission 1960: 31–37.
60. Ibid., 87.
61. Waterston 1963: 98.
62. Ibid., 139.
63. Ibid., 135.
64. Waterston 1963: 33.
65. Ibid., 138.
66. Statistics cannot allow for calculation of real amount of money allocated for transport in the absence of inflation data.
67. Govt. of Planning, National Planning Board 1955: 485.
68. Govt. of Pakistan, Federal Bureau of Statistics 1997a: 4–30.
69. Thorner 1955.
70. Mahoney 2001: 113.
71. LDA 1997: 6–5.
72. Figure presented by director-general, Karachi Mass Transit Cell, City District Government Karachi in the annual seminar of the Chartered Institute of Logistics and Transport, Pakistan in 2004.
73. Mahoney 2001: 113.
74. Waterston 1963: 33.
75. Hasan 1999: 60.
76. Waterston 1963: 33.
77. Hasan 1999: 62.
78. Information gathered from different articles in the journal *Pakistan Development Review*.
79. Ibid., 64, 153.
80. Waterston 1963: 34.
81. Ibid., 99.
82. Noman 1997: 40.

83. Hasan 1999: 103.
84. Ibid.
85. *The New York Times*, 18 January 1965.
86. *The Times* (London), 26 February 1966.
87. Hamid Sultan, *Nawa-i-waqt*, 27 April 2005.
88. Hasan 1999: 51.
89. Hassan 1997: 360.
90. Ibid., 364.
91. Hasan 1999: 89.
92. In 1971, East Pakistan broke away to become Bangladesh. As a result, the Fourth Five-Year Plan (1970–75) was not implemented as originally prepared until the Fifth Five-Year Plan that was launched in 1978.
93. Husain 1984: 105.
94. Govt. of Pakistan, Planning Commission 2001: 18.
95. Easterly 2001: 2.
96. Ibid., 1.
97. Fatima 2003.
98. Hasan 1998: 156–9; Fatima 2003.
99. Hasan 1998: 159.
100. Fatima 2003.
101. Ibid.
102. Ibid.
103. Huda 1968: 340.
104. Hasan 1999: 93.
105. Ibid.
106. Szyliowicz 2004: 23–44.
107. Ibid., 27.
108. Hasan 1999: 37.
109. Non-project aid includes aid for balance of payments, institutional support and relief assistance.
110. Ercelan 2004: 4.
111. Ibid.
112. Hasan 1999: 62.
113. Ibid., 62.
114. Haq 1973: 10.
115. Ibid., 18.
116. Qureshi 1984: 203.
117. Ibid., 337.

4

Pakistan's Transport Policies in the Sustainable Development Era

To answer the question if policies have a self-reinforcement mechanism, one would need to determine if the road-dominated transport policies adopted during the critical juncture (1955–65) remained stable or did they undergo changes, marginal or otherwise. This chapter examines the main transport policy outcomes of Pakistan during the period 1991–2005, made possible by the federal government organizations, international development institutions, transport related professional institutions and the collaboration of all sectors involved in transport decision-making. The year 1991 is selected for two reasons: first, it represents the year when initial efforts were made to formulate transport policy and recorded in documents; secondly, it also marks a period during which the National Conservation Strategy (Pakistan Local Agenda 21) was under preparation. This timeframe will help to assess whether, following the debate on sustainable development, there was any change in road-based transport policies. This chapter is particularly concerned with exploring how the transport policy path was conceptualized and documented so as to provide an overview of the dominant paradigm of Pakistan transport planning during that period. It will also chart environmental issues and show how these were adjusted in transport storylines. Policy output of different institutions will be presented to highlight their approach on transport and related environmental issues at specific times. The review was made through careful examination of transport policies and programmes and sustainable development policies launched during 1991–2005.

Prior to the review of recent transport policies, the chapter describes North's 'interdependent web of institutional matrix' for developing national transport planning and policies in Pakistan.[1] This will be explained in terms of the institutional structure and relationship amongst different hierarchies of organizations in transport decision-making in Pakistan. The review of transport policies and the decision-making process provides a context for the later discussion on the self-reinforcement mechanism—a compulsory component for examining whether path dependence existed in transport planning in Pakistan.

TRANSPORT DECISION-MAKING IN PAKISTAN

Pakistan is characterized by its centralized system of governance in which the federal government plays a dominant role in formulating and implementing policies, including those on transport. This dominance is sustained by virtue of the control it exercises over domestic funds, inclusive of its foreign exchange component, and also the large amount of foreign aid given to the provincial and local governments (see Figure 4.1). Many federal, provincial and local government departments and ministries deal with the transport sector in varying capacities, some in a

FIGURE 4.1
Transport Decision-Making Process in Pakistan

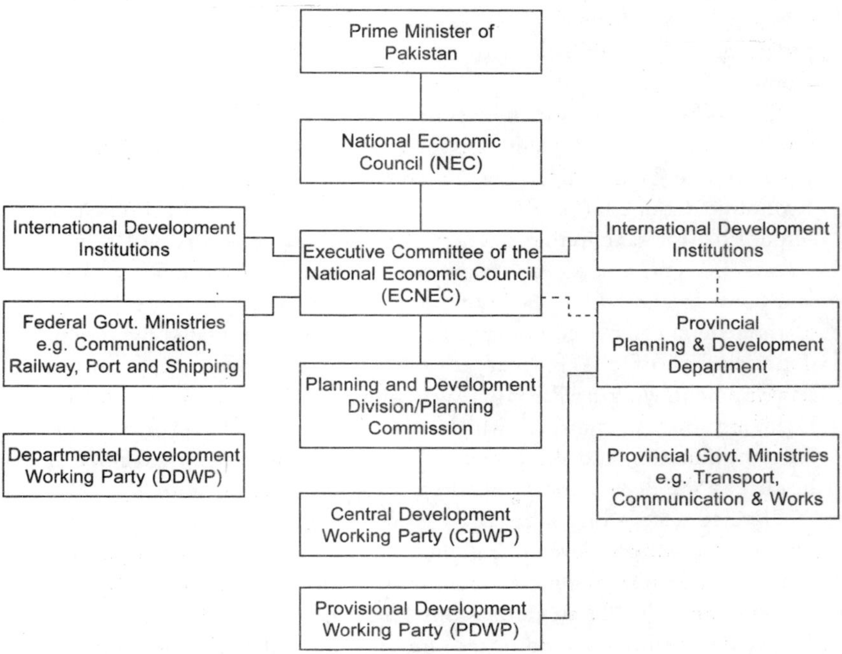

Source: Extracted from Govt. if Pakistan, P&D Division, 1997

comprehensive manner and others only with specific aspects. Figure 4.2 shows the hierarchy of all the organizations in Pakistan at the federal, provincial and local levels who are responsible, either directly or indirectly, for the transport sector.

The Planning Commission and the ministry of finance (shown in Figure 4.2) have no direct link with the transport sector, yet they impact the sector because they prepare and approve budgetary allocations for all major development plans. The Planning Commission, which was established at the federal level in 1958 to replace the Development and Planning Board, is responsible for formulating the long-term perspective plan, the medium-term five-year plan and the short-term annual development plans for the country's economic and social development. In addition, the Planning Commission controls official 'development assistance' from international and bilateral development agencies. Various federal transport ministries and all autonomous organizations under their control, discussed later in the section, are responsible for preparing programmes and projects for their respective modes of transport. Programmes prepared by all federal ministries including transport are submitted to the Planning Commission which coordinates all development programmes and accords approval through the National Economic Council (NEC).

The NEC, a supreme policy making body headed by the prime minister, approves all plans and policies relating to development in Pakistan (see Figure 4.1). The Executive Committee of the National Economic Council (ECNEC) headed by the minister of finance is responsible for sanctioning development schemes in public and private sectors. ECNEC also supervises the implementation of economic policies laid down by the NEC. The Economic Coordination Committee (ECC) of the federal cabinet coordinates economic policies initiated by various departments of the federal government. Moreover, the Central Development Working Party (CDWP) approves all development projects including those in transport which exceed the financial authority of the federal ministries, and the provincial and local governments. CDWP is headed by the deputy chairman of the Planning Commission.

Development projects based on guidelines provided by the five-year plan or the annual development plan are implemented as part of the Public Sector Development Programme (PSDP) formulated by the planning and development division. PSDP is considered by the inter-ministerial priorities committee and the annual plan coordination committee (APCC). When this is cleared by the APCC, the planning and

FIGURE 4.2
Organisations Responsible for Transport Decisions in Pakistan

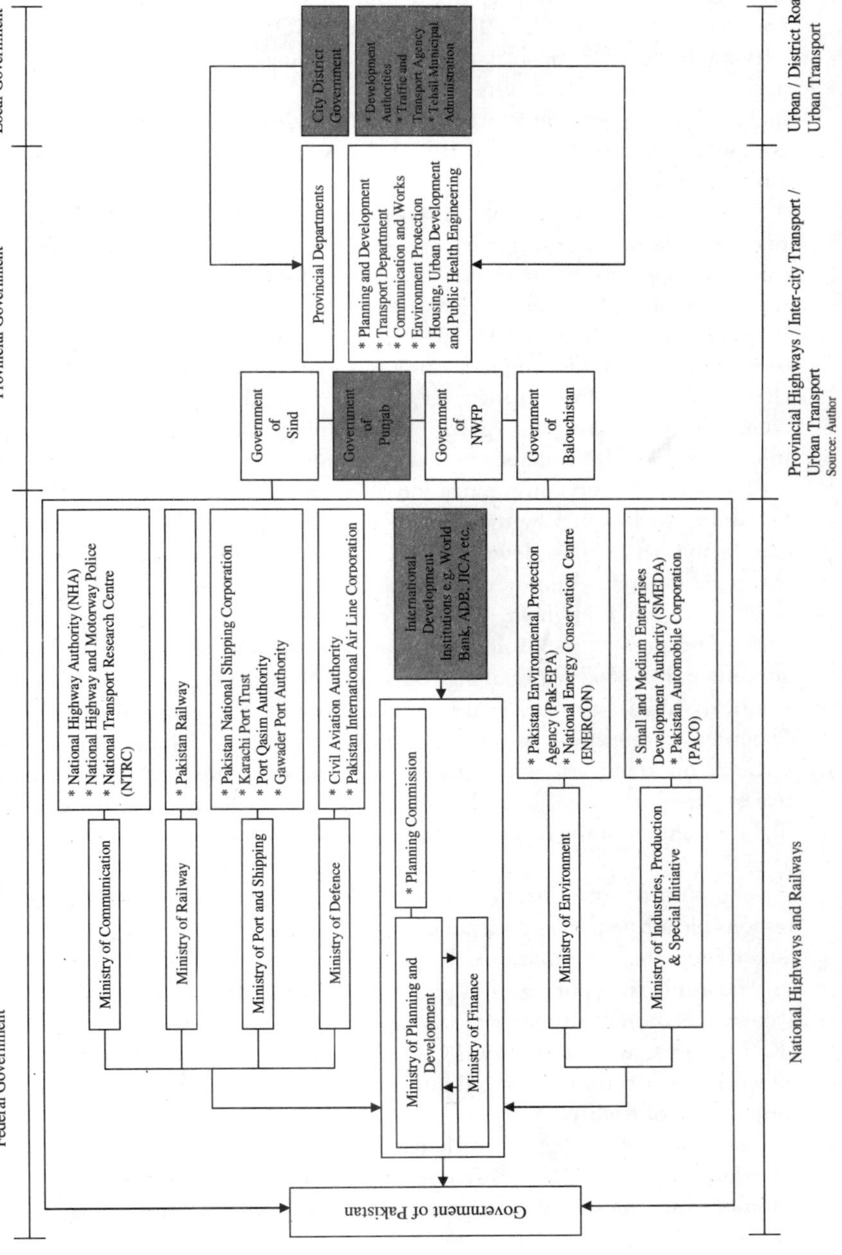

Source: Author

development division prepares a detailed draft of the PSDP for consideration of the NEC. Thereafter, the PSDP is integrated with the national budget for consideration by the National Assembly. PSDP is an annual document which lists all public sector projects and programmes including transport, and allocates funds inclusive of the foreign assistance component for that financial year; it is the operational side of the five-year and annual plans.

From amongst the federal transport ministries and departments, some organizations may be categorized as agencies responsible for infrastructure, others for operation of transport, some others for addressing cross-sectoral issues related to the transport sector and many more that are responsible in some way or the other for all the above mentioned functions. The direct responsibility for implementing transport sector policies falls under the purview of four ministries at the federal government level: communication, railways, ports and shipping and defence. All four ministries deal with different modes of transport.

The ministry of communication is responsible for policy matters pertaining to national highways. A National Highway Council (NHC) was established to lay down policies and guidelines for development, maintenance and operation of the national road network.

The National Highway Authority (NHA) under the ministry of communication is responsible for the construction, maintenance and administration of all national highways. The National Transport Research Centre (NTRC), also under the same ministry but formerly under the Planning Commission, is the sole entity in Pakistan which conducts research on transport, with special emphasis on roads and associated issues. To regulate traffic and enforce rules on highways, the national highway and motorway police was established under this ministry in 1997.

The ministry of railways in conjunction with the Railway Board is responsible for preparing policy and for the control of Pakistan Railways—an autonomous body that operates the railway network in the country.

The ministry of ports and shipping was created relatively recently. The Pakistan National Shipping Corporation (PNSC), Karachi Port Trust (KPT), Port Qasim Authority (PQA) and Gwadar Port Authority (GPA) all work under this ministry and implement its policies in their respective areas of responsibility.

The ministry of defence with its subsidiary organization, the Civil Aviation Authority (CAA), is responsible for air transport in Pakistan. In addition, the ministry of local government and rural development assists the provincial governments in implementing certain rural road projects and programmes.

There are other federal ministries and departments which deal with specific aspects of transport: the ministry of environment and the ministry of industries, production and special initiatives are the two key ministries that fall in this category.

The ministry of environment is the focal point for formulating the national environment policy inclusive of plans and programmes to control transport-related air and noise pollution. The Pakistan Environment Protection Authority (Pak-EPA) is the key organization at the federal level responsible for enforcing Pakistan's Environmental Protection Act, 1997. The National Energy Conservation Centre (ENERCON) works under the ministry of environment to serve as a centre for all energy conservation activities, including transport sector energy use.

The ministry of industries, production and special initiative was established to act as a policy formulating agency for the promotion and expansion of industry by attracting foreign direct investment in Pakistan. The ministry is facilitating the car manufacturing industry as part of a strategy to promote rapid economic development in the country. The Small and Medium Enterprise Development Authority (SMEDA) was established in 1998, also under this ministry, to provide a conducive and facilitating environment for small and medium enterprises (SMEs). Later, SMEDA developed a national transport strategy to attract foreign investment in public transport. Pakistan Automobile Corporation (PACO), which also works under this ministry, was originally created to provide a platform for the coordination and supervision of nationalized automobile companies. Owing to the subsequent privatization of PACO's subsidiary companies, its main responsibility now is restricted to acquiring foreign technology and investment for the automobile sector in Pakistan.

Despite the dominant role of federal ministries and departments, there are also several provincial and local government departments which are responsible for transport infrastructure and operation in their own jurisdiction. Apart from the planning and development department (P&D), the transport, communication and works (C&W), environment protection, housing, urban development and public health engineering departments of the provincial governments are directly or indirectly responsible for transport planning and infrastructure provision in the urban and regional areas of their respective provinces.

In pursuance of the Local Government Ordinance-2001, city district government (CDG), district government (CG), *tehsil*[2] municipal administration (TMA) and union administration have been created in all mega and metropolitan cities, and in urban and rural areas, to decentralize the administrative and financial authority to the local government. In urban areas it is primarily the CDGs who have a significant role in transport and infrastructure policy, planning and implementation. The details of provincial and local government transport organizations are discussed in Chapter 5.

In summary, nine federal ministries and twenty-two departments at all three levels of government contribute directly or indirectly to the transport related decision-making process. By and large it is a very centralized system in which the federal government organizations exercise most power, more so because they have custodial control over both local and foreign financial resources.

PAKISTAN NATIONAL TRANSPORT POLICIES 1991–2005

This section will analyse Pakistan's national transport policy documents for the period 1991–2005 formulated by the federal government departments, international development institutions and transport related institutions, and also trace the collaboration provided by all other sectors involved in transport decision-making. This time frame is greater than the ten years considered necessary to assess the dynamics of policy change.[3] On one hand it is suitable to examine the changing social and economic conditions which provide the constituency base for a major policy change. On the other hand, it provides for the time considered necessary to learn how to formulate policy through a process of technical debates over critical policy issues.[4] This time frame will also help to establish if any changes in transport policy were effected in consequence of the debate on sustainable development.

These policy documents may have promised much, but little was actually delivered as is often the case in the developing world. However, they do reflect the priorities of government and international development institutions during that time period from which one can reasonably infer the policy path. The analysis will identify whether the National Conservation Strategy in 1992 opened a 'policy window' for decision-makers so that they might adopt a fresh approach in addressing the transport problem. In addition, one local case study will be presented in Chapters 6 and 7 to demonstrate what kind of environmental arguments are gaining influence throughout this period. The following section will discuss different transport policy documents formulated during 1991–2005 with particular emphasis placed on their approach to environmental issues.

National Transport Research Centre Transport Policy 1991 and 1992

In the early 1990s, the federal government assigned the task of formulating a national transport policy to the NTRC. The draft document 'National Transport Policy' was published in 1991 and 1992. It covered all modes of transport including railways, roads (including motorways), road transport, urban transport, ports, shipping, air transport and inland water transport. The financial, institutional and environmental aspects of transport were also given due attention in this policy document.

The policy was developed by different working groups assisted by technical experts. This policy-making process was adopted because of the belief that, 'transport is a highly technical field that requires a panel of experts with long standing experience' who can review and formulate policy 'on sound scientific lines'.[5] Along with the need to have technical validation there was also the view that a 'strong scientific bias transport policy should [be] tempered with political realities and popular public feelings'.[6] In pursuance of these objectives, arrangements were made for public consultation (but not public participation).

Although the document states that the public at large was invited through announcements in newspapers to send suggestions for the transport policy, it was found that only ten days were given for this and that the advertisements had been placed in an English-language newspaper with very limited circulation. In a country where little more than 4 per cent of the people have a tertiary qualification, this type of public consultation cannot represent the full spirit of public involvement.[7]

A main concern of this transport policy was to boost economic development by building more roads. Emphasis was given to meeting a road density standard of 0.5 km of road length per sq km of land. This standard was assumed to be necessary for economic prosperity. It was assumed that building more roads and achieving a high level of motorization would bring economic prosperity. Importantly, however, the policy document also noted that despite twenty years of being in operation, the Karachi–Hyderabad superhighway had had no effect on economic development.[8]

Policy documents of this period show a clear bias for adopting bus-based public transport in preference to a rail-based mass transit system as a model for urban transport that would serve the metropolitan cities of Pakistan. The broad change of approach may have been principally driven by the lack of financing from international development institutions to implement a rail-based mass transit system. It was also proposed that government responsibility should be limited to low income groups through provision of an adequate quantum of subsidized public transport. On the other hand, NTRC policy documents proposed that the government should encourage the private sector to provide decent and high quality public transport services for the middle class. Several steps were proposed to encourage the private sector which included grant of soft loans from the banks, and a reduction in custom duty and provision of tax incentives on import of vehicles and their spare parts.

The three academics from UET Lahore (a professor of transport planning, the dean of architecture and planning, and a professor of environmental planning) who were interviewed during field work argued that induction of low quality government-owned public transport in urban areas would project a negative image, ultimately discouraging efforts towards promoting sustainable transport. Similarly, eliminating para-transit such as rickshaws and *tongas* from Pakistani cities, as proposed by the transport policy document, might help in improving the image but it would simultaneously deny cheap mobility to the urban poor, who according to the UN Habitat report 2004 comprise more than half the population of Pakistani cities.

Environmental issues were emphasized but only in the context of how they influence selection of the right option from amongst various proposals for road construction. Thus, technical proposals that tended to improve environmental conditions in cities were preferred. These proposals comprised the use of unleaded petrol, the introduction of catalytic converters, enforcement of existing environmental laws and mandatory conduct of EIA for new development projects. Such priorities revealed a general tendency to conceptualize environmental issues as health and aesthetic issues, and failed to recognize the impact of road construction itself on major ecological problems such as climate change.

Pakistan Motorway Project and Prime Minister's Public Transport Scheme in 1991

The Nawaz Sharif government in the early 1990s attained a degree of fame for introducing transport mega projects in Pakistan. The most important of these were the construction of the Pakistan Motorway and The Prime Minister's Incentives Scheme to Revamp the Public Transport System.

The Pakistan Motorway was planned as a limited access network of 2250 km at a proposed cost of Rs150 billion in 1991. The original idea of the motorway was proposed by a foreign consortium in 1984 to provide additional capacity for the intercity road network between Karachi and Peshawar. A six-lane, limited access tolled Lahore–Islamabad section (333 km) of the motorway with a design speed of 120 km/hr was opened for traffic in 1997. Daewoo Corporation of Korea was awarded the contract on a build, operate and transfer (BOT) basis at an estimated cost of Rs30.5 billion. However, Daewoo funded only 40 per cent of the cost while the remaining 60 per cent was provided by Pakistan. The Lahore Bypass (one section of the Ring Road) was also made to connect the city with the Lahore–Islamabad Motorway section.

The project was justified by the familiar argument that it would lessen congestion, save time and fuel, allow safe and fast travel and reduce vehicle operating costs. On the basis of these justifications, the government projected in 1998 that the Lahore–Islamabad Motorway would generate net revenue of Rs125,000 million in 2026. Additionally, it was envisaged that the motorway would provide transit trade facilities to the Central Asian Republics of the former Soviet Union.

An EIA study was not conducted prior to the construction of the Motorway which was necessary to fulfil the requirement of the Pakistan Environmental Protection Ordinance, 1983. However, massive roadside plantation was planned to improve the aesthetics and purportedly off set the pollution caused by vehicle emissions.[9]

After initiating the motorway project, the Sharif government introduced a policy to revamp the public transport system throughout the country. This policy introduced a number of incentives which included waiving custom duty on import of taxis, buses and mini-buses, granting loans from nationalized banks at 15 per cent annual interest rate for financing these imports and issuing special registration numbers for new public transport. As a result, Pakistan's public transport fleet was upgraded, but after the Sharif government lost office this policy was abandoned.[10]

In brief, the Sharif rule in the early 1990s brought the focus onto road projects and introduced a new fleet of road public transport. Although public transport incentives were withdrawn after remaining in force for two years, work on the remaining sections of the motorway continued with financing from the PSDP. Soon afterwards, the Pindi Bhatian–

Faisalabad section and a partial section of the Islamabad–Peshawar motorway were opened for traffic in 2001 and 2004 respectively.

National Conservation Strategy (Agenda 21) 1992

The National Conservation Strategy (NCS) was the first comprehensive approach to provide a framework for addressing specific environmental concerns of Pakistan. It started as a collaborative exercise in the late 1980s between the government of Pakistan (through its environment and urban affairs division) and the International Union for Conservation of Nature (IUCN). This effort resulted in the issue of a voluminous document; the NCS was formally approved by the government in 1992.

The main objective of the NCS is to facilitate policy for incorporation of environmental considerations in all economic, social and physical development activities.[11] Based on the principle of sustainable development, the NCS explains:

> Sustainability is the key to the NCS approach. It acknowledges that species and ecosystems need to be used for human benefit and that ways must be found to do this in a manner that also ensures their viability and preservation. The values of species and ecosystems become apparent through their contribution to the economic well-being of people, particularly in developing countries. The objective requires avoiding depletion of resources by relying on the interest on natural capital (the annual productive output of renewable natural resources) and on any increase in the quantity and quality of capital stock.[12]

The main message and the explicit objectives of this 'sustainable development' are that the conservation of natural resources and an improved efficiency in their use and management are necessary for sustained development. The NCS categorizes priority concerns into primary, secondary and tertiary sectors. The primary sector is related to agriculture, forest management, rangeland rehabilitation, livestock management, water resources, marine and costal resource management, fisheries, wildlife and their ecosystems and mineral resources.

The secondary and tertiary sectors relate to energy, industrial development, human settlement, pollution control, and recreation and tourism. The NCS identifies fourteen programme areas for priority implementation as a mix of primary, secondary and tertiary sectors.[13] Both the definition and action programmes of sustainable development show clearly the commitment of the NCS towards protection and maintenance of natural resources in the interest of sustainable development. Zahidi explained the reason why the interpretation of sustainable development in Pakistan is limited to an anthropocentric vision.[14] She argues that this is primarily due to the dominant role of IUCN who biased the NCS towards the protection and maintenance of natural resources according to its own vision. Zahidi states that 'the policy making process of sustainable development and implementation exercise of Agenda 21 is under the influence of international development and environmental agencies'.[15] Zahidi's claim is validated by the definition of the sustainable development given in the document.

Transport sectors get very little attention in the NCS perspective. The document recognizes the wider ecological consequences of transport use, and particular attention is paid to the energy and air pollution problem. However, the emphasis is clearly on technical solutions to solve environmental problems associated with improving energy efficiency of motor vehicles. Although the role of public and non-motorized transport in reducing the impact on the environment is acknowledged, at the same time fuel efficient cars are promoted by providing incentives in the form of tax and custom duty relief.[16] There is no concern in the document to reduce motorization or manage traffic growth, and emission control is a low ranking priority. This time the focus was on installing catalytic converters and the use of unleaded fuel.[17] This again highlights the tendency of the document to look for technical solutions to policy problems.

According to Hasan and Ali, there was little political and public involvement during the NCS formulation and hence it became an administrative exercise.[18] They found that institutional capacities of the local government were not compatible with the nature of the proposed projects which makes it hard to implement the NCS. Zahidi found that policies adopted for sustainable urban development in line with the NCS are imported under the influence of the mainstream sustainable development paradigm with the support of international development and environmental agencies.[19] However, the results of these policies remain dependent upon the urban institutional framework. Aina argued that sustainable urban development policies are unsuccessful in Third World cities because their authors lack in-depth knowledge of local institutional capacities. These thoughts are imperative for this study and discussion in the following chapters will describe how the NCS became unsuccessful to make any positive change in the urban areas.[20]

Development Plans in the 1990s

The Second Perspective Plan (1988–2003), prepared by the Planning Commission set the guidelines for the formulation and implementation of the Seventh (1988–93) and Eighth Five-Year Plans (1993–98). As noted earlier, development plans are financed through the PSDP and also by the private sector. The PSDP is financed by federal budgetary allocations including foreign aid and loans channelled through the federal government.[21] The Seventh Five-Year Plan acknowledges that 'Pakistan's ability to formulate and implement the medium-term development plans is dependent on the availability of concessional loans and grants from abroad'.[22] Therefore, the success of the plan is contingent on the government's ability to source funds from international development institutions.

The Eighth Five-Year Plan (1993–98) envisaged a total outlay of Rs1701 billion as compared to Rs660.2 billion for the Seventh Five-Year Plan (1988–93).[23] However, the Seventh Plan estimated that 44.3 per cent of the total money would come from the private sector and 55.7 per cent from the public sector.[24] The Eighth Five-Year Plan reversed this financing and relied more on the private sector.[25] The transport and communication sector gets around 18 per cent of the total plan money in both development plans. The Seventh Plan indicated that the transport and communication sector plan was prepared to use 31.9 per cent of the money allocated: 19.9 per cent from the public sector and 12 per cent from the private sector.[26] It shows that the transport and communication sector secured overall 35.7 per cent of the total PSDP budget in the Seventh Plan. The Eighth Plan figures did not analyse the quantum of PSDP funds used for the transport and communication sector.

In PSDP of the Seventh Five-Year Plan, 12 per cent of the money was allocated to railways as compared to 29.4 per cent to the federal government organizations for road development. Moreover, the Seventh Plan allocated 11.1 per cent of total transport PSDP to provincial governments which again is mainly used for road development. In brief, 40 per cent of transport and communication PSDP was allocated for road development. The rest was for railway, ports, shipping, air transport and the communication sector such as telephone, telegraph and post office. The situation is similar in the Eighth Five-Year Plan where 57.4 per cent of the total transport PSDP was allocated for road development (summarized in Table 4.1).[27] These figures clearly show the government's commitment to road development. Even though every plan tried to set a target to improve the railway and road traffic ratio in favour of the railway because of lack of investment in the railway sector, the railway to road freight traffic ratio fell to 15:85 in 1992–93.[28]

TABLE 4.1
TRANSPORT SECTOR DEVELOPMENT BUDGET IN 7TH AND 8TH FIVE-YEAR PLANS

Development Plans	*Total Development Budget (Billion Rupees)*	*Total Money Allocated for Transport (%)*	*Allocation of Money for Different Transport Sub-sectors during the Public Sector Development Programme (%)*	
			Roads	Others (Railway, Ports and Shipping, Air transport, Telephone, Telegraph and Post)
7th Five-Year Plan (1988-93)	660.2	18	40	60
8th Five-Year Plan (1993-98)	1701	18	57.4	42.6

Source: Figures extracted from Govt. of Pakistan, Planning Commission 1988 and 1994 (7th and 8th Five-Year Plans)

The above discussion shows that the Seventh Five-Year Plan (1988–93) was in fact a road development plan. Certainly the language of the plan provides much evidence of this, for example 'development activities demand mobility...road requires continuous expansion, maintenance, replacement and modernization as an integrated [transport] system[29]... Pakistan has an average road density of 0.146 km of road per sq km of land area which is less than one-third of the generally accepted standard of 0.5 km per sq km of area...the huge deficiency can be made good only through a long term [road] construction programme stretched over many plan periods'.[30] These quotes show the belief system from which stem the justification for promotion of road projects.

An economic growth principle is evident in all road development plans. It is clear that the priorities of the government were economic, as for example 'road development programme...will fill up the gaps in the existing road network with a view to optimally utilizing the system for speedy economic growth'.[31] The road development programme was underpinned by the relationship between traffic growth and GDP. Therefore, the Seventh Plan calculates the 'forecast of the volume and the pattern of the traffic demand for the long and short term based upon projected growth in GDP at 6 per cent'.[32] Moreover, the forecasts estimated an increase in road passenger and freight traffic of between 2.6 and 3 per cent annually by the year 1997–98.[33] The response of the five-year plans to the possibility of more road traffic was to increase the provisioning of infrastructure by perceiving a strong link between transport and economic growth. The road building programme was therefore justified by linking it to Pakistan's economic competitiveness. However, the main principle guiding policy was the government's goal to meet the 'demand for mobility'.[34] Overall, the policy placed more emphasis on mobility than accessibility even though greater mobility for some might often be at the cost of reduced mobility for others. Moreover, both documents attempt to cater for predicted traffic with no evidence that alternatives were considered.

Since the link between road development and economic competitiveness is well established in the minds of decision-makers, they start to find resources for road development. This time the government was looking for funding from the private sector: '...highway programme would be broadened by induction of the private sector and realization of road user charges[35]...introduction of toll road and franchised service centres'.[36] However some emphasis on 'use of existing road network' in parallel to development programmes for broadening highway shows a slight change in thinking but nevertheless all evidence points to a continuity of road development policies.

In terms of environmental and health issues, the Seventh Plan was limited to a dependence on international development institutions. The document states, the 'highway safety programme will be implemented with the World Bank assistance...National Environment Quality Standards (NEQs) will be formulated...all road transport projects will include EIA'.[37] The environment is yet again presented as an un-avoidable by-product of expected traffic growth. Consideration for the environment was presented in such a manner as to keep traffic moving on the roads in pursuit of development. However, the Eighth Five-Year Plan perceived the environmental issue differently because by then the government had formulated the NCS with help from international NGOs. Nevertheless, this time the environment section addressed issues of air quality and energy conservation in the context of transport. It posits that pollution problems should be dealt with by alleviating congestion—building underpasses, bypasses and flyovers. The decade of the 1990s witnessed a realization of this approach, especially in the urban areas of Pakistan. Besides this, the impact of vehicle pollution is dealt with by asking for an increase in engine efficiency. All these measures once again highlight the tendency of the Plans to look for technical fixes and to use environmental slogans for constructing more roads.

The People's Train and Awami (People's) Bus Train Projects

In 1996, under the Prime Minister's (Benazir Bhutto) Development Programme for big cities, a mass transit project was started in the cities of Rawalpindi and Islamabad. This system was based on a rail-road mixed mode which contained an urban rail link between Rawalpindi and Islamabad connected with feeder coasters (small buses) in Islamabad. The main objective of this service was to reduce peak hour traffic congestion, reduce air pollution and make use of existing railway infrastructure.[38]

Initially, the train service was designed for 6000-8000 commuters per day. Therefore, only three train services at the frequency of 1.5 hours in morning peak and three train services at the frequency of 3 hours in afternoon peak was started. However, after three months of operation, these services were reduced to four train services per day. Finally, this rail-road mass transit system was shut down due to heavy financial loss. The main reasons behind its failure was inadequate service planning which included an absence of feeder buses in Rawalpindi, very low frequency, lack of information about timetabling, lack of amenities on railway stations and relatively higher fare without any time saving.[39] Additionally, this train service caused traffic jams on the level crossing roads in Rawalpindi.

A similar kind of project, the Awami (People's) Bus Train was started in 1989 by Benazir Bhutto's first government in Karachi, Rawalpindi and Islamabad.[40] In this project, the National Transport Research Centre (NTRC) had designed and developed a Bus Train (prime mover plus three trailers) using old discarded buses to provide high capacity bus services at peak hours. The Awami Bus Train provided services on the main corridor having sufficient road width. Initially, this project was started in Karachi and after one year of operation, the Bus Train was shifted to Rawalpindi and Islamabad. The Bus Train had introduced an imaginary 'bus lane' on the extreme left of the road for the first time. It was estimated that the Bus Train would attract a large number of commuters in Rawalpindi and Islamabad from 1991 to 1993. This service utilized 45 per cent of its capacity and recovered 68 per cent of its cost from fare in two years of operation.[41] However, this service was also shut down due to lack of interest from the government in providing public transport services.

The Transport Policy of the Chartered Institute of Transport 1998

The Chartered Institute of Transport [now Chartered Institute of Logistic and Transport (CILT)] is an international organization of transport professionals with worldwide presence. In 1998, the Pakistan chapter of this Institute was given a mandate by the ministry of communication to draft a national integrated transport policy and thereby provide a framework for building a modern, integrated transport system that would contribute to Pakistan's progress and prosperity as it prepared to enter the twenty-first century.

The transport policy formulated by the CILT also emphasized use of different modes of transport. A concern of the policy was to address the linkage between land use and transport; it therefore explained that land use should be planned to reduce the need to travel and maximize accessibility of public transport. The policy developed a 'zoning plan' for different land uses, reserving land for future urban transport infrastructure.[42] Even though it is now widely accepted that land use based on separated zones generates more traffic instead of minimizing travel demand, the 'zoning plan' demonstrated that policy makers aimed to reduce congestion by reserving more land for urban transport infrastructure.

The policy envisioned the economical and efficient movement of people and goods in a safe and environmentally sustainable framework. While the policy committed itself to the environment and safety it made no recommendations on how to address these issues. The behaviour of the road users was cited as the root cause of the 'road safety' issue, ignoring the impact of policies favouring motorization and road development. Overall, environmental issues were perceived in a very narrow perspective.

Transport Sector Development Initiative (TSDI) 1999 and NTRC Transport Policy 2001

The Transport Sector Development Initiative (TSDI) was a joint effort between the private sector and the Government of Pakistan in 1999 to collectively develop a comprehensive transportation policy covering all modes of transport. The World Bank made this effort possible by working as a mediator in a series of workshops and seminars arranged to formulate a transport policy. The TSDI policy and recommendations were reproduced by the NTRC in 2001.

The main aim of a comprehensive transport policy is 'to ensure that the infrastructure development and management controls put in place adequately address the interface between each mode of transport'.[43] With this in view, the policy developed a vision:

> a system that provides efficient, safe, reliable, equitable and environmentally friendly access and mobility for people and goods, thereby supporting the governmental goal of increasing public welfare through economic growth, human development and poverty reduction.[44]

Although the vision acknowledged the social and environmental aspect of transport, the policy documents focused on building new roads as a component of economic sustainability in transport development. Certainly this is much in evidence from the language of the document:

> ...it is necessary to increase the length of roads by three times to promote healthy economic activity...construction of new roads should be decided strictly by economic prioritization...economic considerations should govern all decisions on new roads...the road network will provide opportunity to establish small and medium size businesses.[45]

It was also suggested that constructing more roads would help in promoting the automobile industry as that would have the effect of creating a larger market. In short, these documents too favoured a continuation of road building policies.

The TSDI and NTRC policies were framed with a strongly held perspective that privatization and de-regulation of transport and infrastructure would result in more efficient and cost effective development. In relation to privatization, the document states that:

> ...each mode should be developed according to the guidance of market forces....the private sector should be encouraged to play its part in infrastructure development according to the build, operate and transfer (BOT), build, own and operate (BOO) and build, own, operate and transfer (BOOT) basis...a common platform of public and private sector should be established to discuss issues regarding different modes and infrastructure...existing laws and tax duties should be modified in favour of privatization.[46]

An emphasis on privatization was found in all documents prepared with the collaboration of international development institutions in the1990s. The approach towards transport in the above documents was similar to that adopted in the earlier NTRC policy documents.

In relation to environmental issues, the document recommended that 'the government must address health and environmental impacts of transport as a top priority and ensure integration of environmental and economic elements in the project at appraisal stages'.[47] The document did not offer any further explanation on how best to integrate environmental and economic elements into a transport project. It appears that the authors of the document want that proposals for mitigating the negative impact of road construction on the physical environment be made at the project appraisal stage. It makes little sense to suggest that the environmental impact of a road construction project could change the course of its development.

SMEDA National Transport Strategy 1999

After the 1997 general elections, which brought Nawaz Sharif back into office, the government once again gave attention to the transport sector and formulated the national transport strategy; this was developed in 1999 by SMEDA, working under the federal ministry of industries and production. In spite of the fact that several ministries related to transport were in existence, adopting a national transport strategy developed by SMEDA was indicative of the Sharif government's intention to attract private investment in the road transport sector. This strategy was hastily approved by the federal government and a legal notification was issued in May 1999. It should be noted that all previous transport documents remained draft policies which were not formally approved by the government.

Although the strategy was termed the 'National Transport Strategy', it merely focused on the introduction of buses as a mode of urban public transport. Developed under the National (sometimes called the Prime Minister's) Transport Vision, the strategy intended to create 'an integrated, efficient, safe, economically sustainable and market driven quality urban transport system which meets the 21st century needs of citizens of Pakistan, supports rapid economic development and protect the environment'.[48] According to the SMEDA, this 'strategy was envisaged not only to serve the core concept of urban transport but also to help in expediting the economic growth of the country while also addressing environmental issues'.[49]

The main focus of the strategy was to reform government organizations, enabling them to attract private investment in bus-based urban public transport. The strategy therefore attempted to formulate solutions that were customer driven, investor friendly, economically integrated and institutionally focused. As a first step, a National Transport Board (NTB), a federal-level regulatory body, was established under the chairmanship of the prime minister to formulate uniform policies. At the provincial level, it was proposed to establish Provincial Transport Authorities (PTAs) by merging all existing urban transport authorities into a single entity and thus eliminate fragmentation. The PTAs were to be empowered by combining functions such as transport planning, traffic management, infrastructure development, transport regulation and traffic enforcement under a strong, private-sector-dominated board of directors.[50]

The national transport strategy observed that the number of trips made by public transport accounted for only 14 per cent of the total trips in Lahore.[51] This low percentage was stated to be simply on account of non-availability of public transport, which was also described as disorganized and not a respectable form of transport.[52] Since this situation was promoting individual transport modes, causing safety and environmental problems, the national transport strategy introduced franchised bus operations as urban public transport. This was a strategic shift away from new road construction to intensive traffic management through road networking and improved public transport.[53] As a result, favourable policies, tax incentives and regulations were introduced to attract private sector investment in the franchised public transport. Besides this, the government also encouraged commercial institutions and banks to cooperate with private investors in a bid to ensure the success of the franchise system. A significant point the current study discovered is that for the first time in Pakistan a strategy was formulated that overtly discouraged private vehicles as a means of transportation.[54]

In light of its main objectives, the strategy placed emphasis on institutional restructuring to facilitate the flow of private capital. To deliver a quality urban transport system, it suggested that all urban transport related activities at the provincial level—transport planning, traffic management, infrastructure development, transport regulation, vehicles registration and traffic enforcement—be merged and placed under a single (proposed) Provincial Transport Authority (PTA).[55] The emergence of 'investor friendly', 'economically integrated' and 'institutionally focused' storylines was evident in the document. The current study observed the slight change in policy paradigm and the use of appropriate language to justify this change in strategy.

Earlier, the Sharif government had not adopted this strategy at a time when a similar policy had been introduced by the provincial (Punjab) transport department. After five-years of franchise bus operations in various cities of the Punjab, the government notified a few minor modifications to its policy on the system of franchise bus operation, on 29 March 2005.

Planning Commission Transport Policy 2000

In 2000, the federal Planning Commission prepared a draft 'Transport Policy' through an in-house process but which was sent to all provincial transport departments for their views and comments.

The Planning Commission policy also presented a bus-based public transport system as the transport solution for metropolitan cities of Pakistan. This policy was the first to propose reserving special bus lanes in grade or grade-separated road infrastructure. The policy encouraged revitalization of the KCR as an urban rail line, which had been abandoned in the late 1990s. However, in contrast to these recommendations, the policy also suggested eliminating para-transit vehicles. The aim of this recommendation was to provide more road space for cars and public transport, in spite of the fact that para-transit vehicles serve the poor and low income groups in Pakistani cities. As the private sector was encouraged to operate public transport, para-transit remained the only affordable mode of transport for the urban poor.

In relation to environmental issues, the policy presented a series of technical responses such as introduction of CNG or liquid petroleum gas (LPG) and catalytic converters. However the document was ineffective in addressing the full impact of transport development on the environment.

Review of National Conservation Strategy

To assess the achievements, impacts and prospects of the national conservation strategy, a mid-term review (MTR) prepared by foreign consultants was completed in 2000. It reported that the scope of the NCS was limited with regard to raising awareness and building institutions and instead the emphasis was on promoting actual improvements in the quality and productivity of the environment and natural resources. The MTR argued that the NCS was not adequately focused as a national sustainable development strategy, and needed to go beyond conservation objectives by focusing on achieving wider development outcomes.

The review observed that the NCS focused less on urban air pollution, pollution taxes and user charges related to transport.[56] However, it acknowledged the role of the NCS in introducing, promoting and establishing a CNG infrastructure in Pakistan to assist in instituting a more environmentally sustainable transport.[57]

In summary, the MTR concluded that the NCS did not result in an overall policy shift towards sustainable development, with the minor exception of some environmental and resource management policies.[58]

Asian Development Bank (ADB) Transport Policy Document 2003

The latest effort to formulate a national transport policy was initiated at the end of 2003 with the technical assistance of the Asian Development Bank. Stage one of the assistance appeared in the form of 'Assessment of Critical Current Transport Sector Needs', a report prepared by international consultants appointed by the ADB. The report proposed a mission statement and national transport policy guidelines which defined transport priorities and principles for transport organizations in Pakistan. The proposed mission statement was 'to provide the people of Pakistan a safe, reliable, affordable and modern transport network and services to effectively support the economy and act as a catalyst for economic development and social growth'.[59] Transport priorities emanating from this mission statement included, (1) strengthening the regulatory framework of the transport industry (2) supporting continued growth and improvement of the national and provincial highway network, and (3) revitalization of the railways. Priority one stressed the regulatory role of the government and proposed instituting performance indicators that were to be monitored to ensure effective enforcement. The policy suggested establishing a transport council (TC), transport coordination and cooperation (TCC) committees and task forces for sectoral cooperation and coordination.[60] It was proposed that investment in infrastructure be based on network efficiency and poverty mitigation priorities. Stage two of this technical assistance, which will mainly contain sub-sector policy statements, has not yet commenced.

Development Plans in the 2000s

The Ten Year Perspective Development Plan prepared by the Planning Commission is to be implemented between 2001 and 2011. In light of the objectives stated in the Plan a three year development programme was prepared for implementation between 2001 and 2004. Similarly, a Medium Term Development Framework (MTDF) was made for implementation between 2005 and 2010. The Ten Year Perspective Plan identified transport and environmental issues and formulated strategies and operational programmes. An ambitious road development plan was recommended although it was recognized that its implementation would depend on foreign assistance.[61] A draft copy of the MTDF was made available in March 2005 for public consultation. The Planning Commission placed the document on its website although only 1 per cent of the population had access to the Internet in 2002 (According to the UN Common Database, Pakistan ranked 115 out of 125 countries in the world in 2002 for use of the Internet).[62]

The MTDF was prepared by constituting thirty-two working groups with involvement of experts from federal and provincial governments, public and private sectors, civil society organizations, academia and research and (international) development institutions.[63] In the context of Pakistan, experts from international development institutions have historically enjoyed a clear supremacy over others because of their central role in providing technical and financial assistance.

The MTDF (2005–10) was prepared in light of Vision 2030. This vision, following the tradition of meaningless overstatements, envisages a 'developed, industrialized, just and prosperous Pakistan through rapid and sustainable development in a resource constrained economy by employing knowledge inputs'.[64] To achieve this vision, the government claimed that the MTDF introduced a paradigm shift by putting development of infrastructure, human resources and technology together into an integrated plan named 'Pakistan Incorporated'. The paradigm shift seems apparent from this statement made by the former minister of science and technology, Dr Atta-ur-Rehman, during his keynote address at a seminar on the 'Future of Pakistan and Current Higher Education': A 'nation develops not by building roads and bridges but through education and research.'[65] However, the perception of a paradigm shift needs careful assessment in order to evaluate the transport proposals made in these documents.

The picture that emerges in the transportation and communication sectors appears to contradict the aims set in Vision 2030. The principal focus remains constant on building roads as a major component of infrastructure development while only 0.03 per cent of the transport and communication sector PSDP has been allocated for research. Even though the document acknowledges at several places that the nation would not develop without research and education, the financial allocation for this purpose clearly shows the absence of commitment.

The MTDF envisages an investment of Rs2094.2 billion under PSDP, the bulk of which (73.8 per cent) would be made by the federal government mainly in the infrastructure sector. Major components of infrastructure outlays are for energy, transport and communications and water resources. The four provinces in combination obtained only 26.2 per cent of the PSDP budget, to be invested in the social sector. The transport and communication sector secured 15.5 per cent of the PSDP budget for 2005–10. This money excludes Rs563 billion (63.4 per cent of total money allocated for transport and communication) which would be generated under the self-finance programme and public-private sector partnership for transport and communication during the plan period.

It was found that 61.1 per cent of the total transport and communication PSDP allocation was spent on road development during 2001–5 (we have added 80 per cent of the provincial transport budget). The MTDF planned to spend 63.4 per cent of the total PSDP funds on roads, 19.3 per cent on railways and the reminder on ports, shipping, aviation, information technology and telecommunication sub-sectors. This is in addition to the Rs30.9 billion allocated for road development under the public-private financing mechanism. Urban roads and related infrastructure such as flyovers and interchanges have attracted 27 per cent of the money allocated under the public-private partnership programme.

Another hidden resource for road development was discovered. This time road infrastructure was included in the 'Urban Development and Housing' sector of the MTDF. It identified Rs39.7 billion allocated for construction of flyovers, interchanges, overhead bridges, underpasses and ring roads in urban areas, which constitutes 61 per cent of funds allocated in the PDSP for urban development. On the other hand Rs3 billion (4.6 per cent) was allocated for large public transport terminals in major cities of Pakistan such as Karachi, Lahore, Faisalabad and Rawalpindi. No money was allocated for any type of public transport and non-motorized transport in these cities.

The MTDF transport vision initiated the use of the word of 'integration'. For example, the MTDF emphasized the need to develop:

> ...an efficient and well integrated transport system...establishment of integrated urban development approach...comprehensive and integrated transport policy will be developed during the Plan...Mega cities have to... improve integrated development of urban infrastructure...environment have to integrated into all development efforts and policy formulations...integration of social and environmental considerations into development process...is very essential.[66]

In doing so an emphasis had been placed on the role of planning and land use because in the past, the MTDF observed, poor planning had resulted in making a 'graveyard of projects'.[67] It therefore proposed the development of a 'National Urban Policy' by creating an 'Urban Policy Research Centre' in the Planning Commission.[68] Moreover, it asked that a 'National Spatial Strategy' incorporating national and provincial spatial development plans ought to be prepared. At the local level, it pointed out the need to prepare district structure plans and *tehsil* master plans for guiding the growth of urban and rural settlements in the direction of the provincial spatial plan. Action plans for specific local areas were to be formulated in the context of the district structure plans and the *tehsil* master plans.[69] The MTDF proposed to empower these plans by legislating a 'Spatial Planning Act' to permit creation of the necessary agencies to prepare and implement spatial strategies at the national, provincial and local levels.[70]

The MTDF is the first document in which emphasis has been placed on planning. It also highlights the absence of planning land use as an issue for growth of transport. The emphasis at places is on planning for mixed use of land as when it says that, 'a priority will be a...modification of regulation to allow mixed use of land'[71] but elsewhere it wants to 'introduce land use planning and zoning'[72] thereby endorsing the old concept of distinct zoning. These statements raised questions about how the government was expected to deal with transport in the presence of policies that called for both zoning and mixed land use. The concept of zoning was introduced in the era of motorization in which separate zones were established for different land uses. But since experience has shown that creating zones have in reality promoted long journeys, today one must plan for a mixed land use in order to reduce the need for travel.

Solutions to reduce road traffic are very dependent on how these are perceived by the federal government. In the MTDF, there are several contradictory policies especially between those for demand management and provision for demand. On one hand, the document stresses demand management, stating that 'urban transport investment would incorporate the concept of demand management[73]...traffic management schemes will incorporate solution to optimize the efficient use of road space'.[74] On the other hand, the MTDF advocates a 'predict and provide' approach, declaring that 'the performance of transport system has been poor...due to mismatch between supply and demand for transport service and supporting infrastructure[75]...catering to private transport demand is also important as a high proportion of future trip making will continue to be made by cars, motorcycles, rickshaws and bicycles'.[76] These statements clearly highlight the emphasis on demand and allow us to argue that the notion of demand management is a storyline with limited significance.

While the MTDF stated that the 'development of an efficient public transport system primarily based on buses needs to be linked to mass transit systems, with light rail as an option', it also calls for large scale expansion of the road network on the basis of 'primary, secondary and tertiary roads...in Karachi and Lahore.'[77] In addition, non-motorized transport such as bicycles has been included under private motorized transport to obtain higher figures that would justify provision of infrastructure to meet demand. However, data reveals that during the era of rapid growth in the number of cars and motorcycles, the number of non-motorized transport especially bicycles declined. In conclusion, the principal belief underpinning the policy report is that there can be no halt to traffic growth. The real solution, the report concludes, is to make users cover the full cost of the transport infrastructure which may be done either by imposing direct road user charges or by introducing toll payments on more roads.

The discourse on enhancing economic competitiveness by building more roads has been very strong throughout the history of Pakistan. As the MTDF states,

> economic development cannot take place without the availability of proper infrastructural facilities...an efficient transport and communication system contributed to economic growth[78] action will be required to provide adequate infrastructure to support urban-based economic growth.[79]

A key storyline for infrastructural development lies in the belief that roads will bring economic development. A slight change in the perception of decision-makers was seen during the days following Dr Atta-ur-Rehman's keynote address (mentioned earlier in the section) but it seems that this was just a beginning because the majority of decision-makers still favour road building as a means to spur economic growth. The economic growth argument has always won over environmental concerns.

Environmental measures continued to receive an 'end of pipe' treatment. While the chapter on environment in the MTDF does address the issue of air quality as a rising health problem and therefore an annual cost on the economy, it interprets air pollution as a result of excessive congestion stemming from an inadequate infrastructure rather than a problem caused by growth of traffic. The solution suggested by the documents lies merely in building bypass roads, flyovers and underpasses in order to keep traffic moving. Thus, air quality concerns were used to justify more roads and ironically, rising environmental awareness actually enabled the road sector to obtain more funds in the overall PSDP. This is evidenced in the distribution of PSDP funds which allocated Rs4.56 billion for the environment sector as compared to Rs248.39 billion just for building roads throughout the country during 2005–10. This amount is 11.5 times larger than the total environment sector budget.

The impact of vehicle pollutants on the global climate was mentioned briefly but it was considered that the best solutions were to 'increase engine efficiency, encouragement of CNG fitted vehicles and substituting fuels'.[80] The document also underlines the belief that the environmental situation would improve by introducing regulatory measures, the most noted of these being the strategic environmental assessment (SEA), environmental accounting (EA), environmental information management (EIM) and environmental impact assessment (EIA).[81] But arguably there is still a dominance of traditional approaches and a lack of understanding of the wider ecological issues.

Institutional issues, especially 'highly centralized decision-making' in the national and provincial transport organizations also get attention in the MTDF.[82] For better governance and fast economic growth it has been proposed to decentralize many functions to the local level. This included appointing of provincial finance commissions (PFC), *zila mohtasibs* (ZM) and citizen community boards (CCB). While the MTDF proposes instituting a 'people-centred approach' for service delivery at the local level, it also notes that 'local governments have inadequate technical and financial capacities'[83] and hence the challenge would be 'to build capacities of local administration for integrated urban development activities, particularly in metropolitan cities and mega cities where there is extensive involvement of federal and provincial government agencies'.[84] The present study, however, found that there was an absence of a broader view on capacity building which was restricted mainly to seminars and on-the-job training.

In summary, eighteen policy documents were produced in the last fifteen years which addressed transport directly or indirectly. These documents consistently affirm road building as the main tool for transport development in the country. Environmental concerns have been addressed by measures that were limited to technical fixes in automobiles and some regulatory initiatives. Overall, the transport policy of Pakistan has not undergone any major change following the critical juncture described in the previous chapter.

NOTES

1. North 1990.
2. Pakistan's administrative structure comprises of four provinces, i.e., the Punjab, Balochistan, Sindh and North West Frontier Province and 102 districts. Districts are further subdivided into '*tehsils*', which may contain a number of villages or municipalities. Moreover, the adminsitrative structure also includes Islamabad Capital Territory (ICT), seven Tribal Agencies, Azad Kashmir and Northern Areas.
3. Sabatier 1993: 15.
4. Ibid., 15–16.
5. Govt. of Pakistan, NTRC 1991: 1–5.
6. Ibid., 6.
7. http://www.statpak.gov.pk/depts/pco/statistics/pop_education/pop_education_sex.html
8. Govt. of Pakistan, NTRC 1992: 31.
9. http://www.iepsac.org/papers/p04a11.htm
10. From the brochure 'The Prime Minister's Incentives Scheme to Revamp the Public Transport System' prepared by the Ministry of Communication, Government of Pakistan.
11. Govt. of Pakistan, EUAD and IUCN 1992: 140.
12. Ibid., 141.
13. These are: (1) Maintaining soil in cropland (2) Increasing irrigation efficiency (3) Protecting watersheds (4) Supporting forestry and plantations (5) Restoring rangelands and improving livestock (6) Protecting water bodies and sustaining fisheries (7) Conserving bio-diversity (8) Increasing energy efficiency (9) Developing and deploying renewable sources of energy (10) Preventing/abating pollution (11) Managing urban waste (12) Supporting institutions for common resources (13) Integrating population and environment programmes (14) Preserving the cultural heritage.
14. Zahidi 1998: 122.
15. Ibid., 127.
16. Govt. of Pakistan, EUAD and IUCN 1992: 227.
17. Ibid., 235.
18. Hasan and Ali 1992.
19. Zahidi 1998.
20. Aina 1990.
21. Govt. of Pakistan, Planning Commission 1988: 92.
22. Ibid., 13.
23. Govt. of Pakistan, Planning Commission 1994: iv.
24. Govt. of Pakistan, Planning Commission 1988: 91.
25. Govt. of Pakistan, Planning Commission 1994: iv.

26. Ibid., 541–2.
27. Ibid., 509.
28. Ibid., 264.
29. Ibid., 491.
30. Ibid., 505.
31. Ibid., 268.
32. Ibid., 494.
33. Ibid., 264.
34. Ibid., 263.
35. Ibid., 493.
36. Ibid., 505.
37. Ibid., 510.
38. Govt. of Pakistan, NTRC 1996: 3.
39. Ibid., 8.
40. Govt. of Pakistan, NTRC 1992.
41. Govt. of Pakistan, NTRC 1996: 19.
42. CILT 1998: 11.
43. www.pakroadusers.com/tsdi/index.htm
44. Govt. of Pakistan, NTRC 2001: 5.
45. www.pakroadusers.com/tsdi/index.htm
46. Ibid.
47. Govt. of Pakistan, NTRC 2001: 29.
48. Govt. of Pakistan, SMEDA 1999: 22.
49. Ibid., 21.
50. Ibid., 12.
51. Ibid., 57.
52. Ibid., 73.
53. Ibid., 14.
54. Ibid., 74.
55. Ibid., 80.
56. Hanson et al., 2000: 34.
57. Ibid., 55.
58. Ibid., 95.
59. ADB and Govt. of Pakistan, NTRC 2003: 1–2.
60. Ibid., 1–11.
61. Govt. of Pakistan, Planning Commission 2001: 18.
62. http://globalis.gvu.unu.edu/country.cfm?country=PK&indicatorid=0
63. Govt. of Pakistan, Planning Commission 2005: Preface.
64. Ibid., Introduction.
65. *Daily Jang*, 30 March 2005.
66. Ibid., Transport and Environment sections.
67. Ibid., Developed Development, 2.
68. Ibid., Urban Development and Housing, 9.
69. Ibid., 9.
70. Ibid., 10.
71. Ibid., 8.
72. Ibid., Environment, 10.
73. Ibid., Urban Development and Housing, 6.
74. Ibid., 9.
75. Ibid., Transport, 2.

76. Ibid., Urban Development and Housing, 8.
77. Ibid., 8.
78. Ibid., an Overview, 9.
79. Ibid., 1.
80. Ibid., Environment, 7–10.
81. Ibid.
82. Ibid., Transport, 5.
83. Ibid., Urban Development and Housing, 1.
84. Ibid., 2.

5

Exploring the Stability of Road-based Transport Policies in Pakistan

Pakistan's transport policies during the period 1991–2005 show signs of a slight easing of the hold that the road-based transport policy paradigm had established in the post-critical juncture (1955–65) period. Environmental concerns emerged in the early part of this period but by incorporating a narrow definition of the environment, transport policy continued to retain road construction as the main policy instrument. As a whole, an analysis of various transport policies introduced over an extended time period will help us to determine their stability. This chapter will explore factors that helped in generating the self-reinforcement mechanism which stabilized road-based transport policies initiated during the critical juncture (1955–65). The discussion will explain how the transport planning paradigm in Pakistan has become increasingly difficult to change.

It was evident in the previous chapters that Pakistan relies heavily on financial and technical assistance from international institutions. This chapter will explore the power these institutions exercise over transport planning in Pakistan by identifying weaknesses in the structure and capacity of local transport institutions. Institutional analysis will lead the discussion into exploring the process of justification for transport solutions adopted over time. Discursive analysis will follow in which this study will explore how decision-makers perceived transport and environment, constructing new storylines and promoting the dominant transport planning paradigm. In addition, given that transport infrastructure and operations are ultimately delivered to specific places, the local case study presented in Chapters 6–7 provides the fine-grained analysis that indicates which institutions and storylines gain influence throughout this period.

INSTITUTIONAL AND POWER RELATIONSHIPS AMONG INTERNATIONAL AND NATIONAL INSTITUTIONS

A study of all the various transport policy documents, juxtaposed with the periods and the sequence in which they were launched over the last fifteen years help us to understand the reasoning behind Pakistan's current transport policies. Besides, it would also help us to identify the principal paradigm which guided Pakistan's transport policy during this period in which a total of eighteen transport policy documents were formulated by national and international institutions. However, most of these documents failed to adopt a holistic approach and instead prioritized road building as central to solutions that would address the problem of transport in the country. Although the significance of the environment was recognized, the full impact of road building on the environment was not given due consideration. The policy was limited to advancing technical solutions for mitigating the environmental impact of rapid motorization which itself resulted from a transport development strategy that laid emphasis on road building. In light of the discussion in the previous chapter, it is obvious that stability or continuity has clearly been a dominant characteristic of transport policies over a long period of time. Does this stability result in creating resistance to change in the initial transport policies, more so in the presence of a national and international environmental agenda of the 1990s? Returning to the theoretical framework developed for this study, it was suggested that policies often generate self-reinforcing mechanisms from beyond the critical juncture period. This section will explore the institutional factors responsible for generating self-reinforcing mechanisms that stabilized transport policies over time. The fact that these factors evolved over time and appear to help in resisting change to the pre-established pattern of transport policies in the country will be also be discussed in the following sections.

EXPLORING THE INSTITUTIONAL CAPACITY OF NATIONAL INSTITUTIONS

It was determined that in the early 1950s there was a lack of capacity for transport policy development in all federal, provincial and local government organizations. This serious deficit in capacity compelled the government to invite international development institutions and their associated foreign consultants to prepare the First and Second Five-Year Plans (1955–65). One of the tasks assigned to them at that time was to develop local institutional capacities and enable them to conduct future research and planning without outside assistance.[1] How and to what extent this was achieved, especially in transport related institutions will be explored which in turn will help in establishing the nature of power relationship amongst local and international institutions. Sustainable transport infrastructure and policies depend critically on a country's institutions and their capacities.[2] According to Grindle, 'a capable state is one that exhibits the ability to establish and maintain effective institutional, technical, administrative and political functions'.[3] Jenkins and Smith have employed institutional, political, administrative, technical and economic capacity analysis in analysing housing systems in South Africa and Costa Rica.[4] In Pakistan, the National Environmental Action Plan (NEAP) (2001) has broadly defined 'capacities' in terms of substantive capacities (policy formulation, institutions, human resources and financial) and support capacities, (research and information). The current study uses these categories of capacities for exploring existing transport institutions in Pakistan.

a) Policy and Legislation Formulation Capacity

Pakistan was one of the pioneering countries in the developing world which instituted deliberate public policies on transport. These policies were part of a development planning exercise which started with the preparation of the First Five-Year (Development) Plan in 1955. From then up to the recently approved MTDF (2005–10), urban transport programmes were formulated under 'Transport' as a distinct sector head. In the early 1990s and 2000s, several efforts were made by the NTRC, CITP, SMEDA, World Bank, Planning Commission and the ADB to formulate national transport policies. Although none of the policy documents prepared by these agencies were formally approved, many concepts contained therein have become the *de facto* policy of sub-sector agencies.

In terms of legislations, Pakistan promulgated the Motor Vehicle Ordinance 1965 and the Motor Vehicle Rules 1969 aimed at controlling air and noise pollution from vehicles. In another step, the Punjab Cities and Development Act, 1976 established 'development authorities' in metropolitan cities for comprehensive development and urban transport planning. However, the first substantive effort linking the environment with urban transport was the promulgation of the Pakistan Environmental Protection Ordinance (PEPO) 1983. Pakistan's National Conservation Strategy (NCS) 1992 provided overall guidelines for sectoral policies and programmes for sustainable development. The National Environmental Quality Standards (NEQS) were established in 1993 setting benchmarks for industrial pollution as well as standards for smoke and noise emissions from motor vehicles. PEPO was replaced in 1997 by the Pakistan Environmental Protection Act (PEPA) that empowers environmental institutions to formulate environmental policies and establish funds for sustainable development. However, local academics found a 'lot of weakness in PEPA, 1997 for dealing with transport pollution in the absence of appropriate standards' (Associate Prof. of Environmental Planning, UET Lahore). In addition to all these, the National Highway Safety Ordinance 2000 was enacted to provide for safe driving on national highways. In summary, all these acts, legislations and policies have put in place programmes reflecting some components of sustainable development and transport.

b) Institutional Capacity

In Pakistan, it is primarily the federal government that deals with transport with little participation of the provincial and local governments. However, both the federal and provincial governments have a number of ministries and divisions which are directly or indirectly responsible for transport. Consequently, transport policies reflect this highly inefficient and fragmented structure comprising of some eight federal ministries plus fourteen departments, seven provincial ministries plus six departments and two local government departments—all with overlapping responsibilities and no coordination.[5] There is consensus among national experts and researchers that there is a lack of coordination amongst the various transport institutions in Pakistan. For example:

> ...lack of coordination at all levels, amongst different wings within the same organization, inter-ministerial coordination and from federal to provincial and local level is one of the main reasons for poor performance of transport institutions (Deputy Chief, Physical Planning Section, Planning Commission).
>
> ...federal and provincial institutions have lack of coordination for environment protection. Pak-EPA has almost no contact with Regional Transport Authorities (RTA) on matters relating to pollution and transport (Director, Pakistan Environmental Protection Agency).
>
> ...there is no coordination mechanism defined in the hierarchy of organizations. In any project, the process of decision-making, the specific role of each organization and the responsibilities of various particular are not defined in Pakistan (Programme Officer, Pakistan Planning and Management Institute).

Owing to this lack of coordination, most institutions at different levels in the hierarchy work within their own departmental and professional boundaries. Also, since the subject of transport is dealt with by different departments and relates to several areas of expertise, it becomes even more difficult to coordinate and get a 'joined-up-thinking'.

A gradual change in the federal government's approach to transport and infrastructure planning has been taking place since 1947. In the early 1950s, government organizations were responsible for providing public transport services and for building transport infrastructure. This policy was altered in the 1960s when the government created corporations (semi-public) to provide urban public transport services. After the condition of these corporations deteriorated in the late 1980s, the private sector was encouraged to invest in operating public transport in parallel with low-quality, government-owned public transport services. In the early 1990s, there were major changes signified by the thinking that government organizations had only a limited role in providing transport infrastructure and services, and that the private sector had a responsibility to invest in public transport operations (see TSDI and NTRC 2001 and ADB and NTRC 2003 transport policies). In conjunction with this, the government accepted the role of promoting a more favourable climate for private sector investment in the transport sector.[6] Some of these policies were designed to reduce risk, some to speed up delivery and some to change attitudes to transport investment. The motorway project (high-speed, intercity road network that started in the early 1990s) and foreign investment in franchised public transport services in various cities of Pakistan are the outcome of these polices. However, the legitimacy of these policies has been weakened because no alternative institutional structure and mechanism, such as new legislation to achieve the goal of wider private sector investment, has been created to replace the old process of transport infrastructure supply.

In his institutional study for the city of Lahore, Meakin identified the fragmentation of urban transport responsibilities, the absence of effective coordination mechanisms, lack of monitoring and evaluation and an approach based on solving urgent day-to-day problems amongst several government transport organizations.[7] Husain pointed out that political instability, over-centralized power in favour of the federal government, and lack of institutional capacity—such as the absence of clear and coherent goals, lack of financial and managerial autonomy and no accountability—were the main weaknesses of infrastructure planning institutions in Pakistan.[8] Husain concluded that these institutional weaknesses were giving rise to an 'elitist growth model' in the overall development paradigm in Pakistan.[9]

c) Human Resource and Technological Capacity

Pakistan did not come into being as an independent state with a blank slate. It inherited a system of government highly dependent on a civil bureaucracy but one that was primarily designed to attain the colonial objectives of regulating behaviour, extracting resources, and delivering fair and impartial administration; it was not designed to promote development activities.[10] For promoting development activities after independence, Pakistan's Planning Commission enrolled foreign advisors from the Harvard University to prepare the First and Second Five-Year Plans covering the period from 1955 to 1965. One objective of the Harvard Group was to train staff of a central planning body in Pakistan to carry out research and planning activities without outside assistance,[11] but this was not attained.[12] A local researcher has also pointed this out:

> ...Building capacity of the institutions was totally neglected by the Harvard group in the late 1950s and early 1960s. As a result only three people are currently employed for Physical Planning, three for Transport, three for Environment and one for the Macro Modelling Section in the federal Planning Commission, which is the highest planning institution in Pakistan. These few people are responsible for developing country models for their respective sections (Programme Officer, Pakistan Planning and Management Institute).

Mahnaz Fatima has argued that the initial lack of human resource capacity in planning institutions in Pakistan resulted in increasing intellectual dependence over time on external sources of advice.[13]

The Planning Commission did employ a few professionals for transport planning and engineering but on the other hand, a 'planning unit' does not exist in most federal transport ministries. The First Five-Year Plan (1955–60) recommendation to establish a 'planning cell' in each ministry has therefore been totally ignored. Overall there is a 'lack of proper institutional framework and human resource capacity for undertaking (transport) planning' at the national level.[14] For example the NHA continues to be dependent on external foreign consultants on whose advice some major road projects have been launched in Pakistan which are larger in size and specifications than is warranted by local traffic and geo-technical conditions.[15] Moreover, environmental institutions that were established in the last decade also suffer from a lack of human resource capacity as pointed out by a senior expert of the federal government:

> ...environmental institutions in Pakistan are at the initial stages. Our financial and human resource capacity is very weak. Our performance is as you would expect, keeping in mind that we have only three to five environmental engineers working at the national level. According to the policy, every ministry should have an environmental manager, but in practice no one has been employed there (Director, Pakistan Environmental Protection Agency).

The situation is even more dismal in organizations of the provincial and local governments, as will be discussed in Chapter 6. In summary, it is quite evident that Pakistan's national institutions which deal with transport and the environment are severely handicapped on account of serious deficiencies in human resources.

d) Financial Capacity

Transport institutions in Pakistan aim to achieve a self-sustaining financial capacity. All other institutional capacities would not be enough to provide sustainable transport if matching funds are not made available. There is a consensus amongst experts that 'the major barrier to development in Pakistan is the non-availability of required funds' (Deputy Chief, Physical Planning Section, Planning Commission).

Historically, it is the government institutions which provide finance for the transport infrastructure in the country. Urban transport policies and programmes are the responsibility of the provincial governments who work in collaboration with each city government. The federal government has no formal, constitutional control over planning agencies or development activities in the provinces and in cities. In practice, however, city and provincial governments depend on the federal government for funds to finance most of their transport programmes.[16] By virtue of the control that the federal government exercises over financial resources, it has attained a position of dominance *vis-à-vis* the provincial and city governments.[17] The federal government collects 92 per cent of the gross revenues, and the provincial and city governments just get the remaining 8 per cent. This shows that the revenue base of each province and city government is very limited and sub-national governments are heavily dependent on federal government grants.[18] However, the federal government itself is always short of financial resources to sponsor transport infrastructure development in major cities of Pakistan since these require substantial funds.

As a result, the federal government relies on international development institutions to finance development of the transport infrastructure. Husain found that foreign finance comprised 80 to 90 per cent of development expenditure in Pakistan.[19] Burki demonstrated that Pakistan's reliance on external finance to meet its development expenditure increased over time.[20] The following quote describes how one expert views the financial dependence of the country:

> ...non-availability of finance is the main hindrance for transport development in Pakistan. Financing can be made possible through private sector participation, e.g. BOT, foreign aid and local finance. Throughout history, only funds obtained through foreign aid or loans were made available for transport development in Pakistan (Chief, Roads and Bridges, P&D, Govt. of Punjab).

In 2005, the transport sector was allocated the lion's share of 25.7 per cent of the total loan and credit given to Pakistan by the World Bank yet development in this sector was mainly restricted to the national highway rehabilitation project. The transport sector was allocated the lion's share of 25.7 per cent of the total loan and credit given to Pakistan yet development in this sector was mainly restricted to the national highway rehabilitation project. Similarly, the Asian Development Bank's financial assistance to Pakistan as of 2004 allocated 18.4 per cent of the money for the transport and communication sector; financial assistance for the transport sector was committed to rehabilitating, developing and integrating the road network, enhancing capacity for better road assets management, operation and maintenance, and for institutional strengthening of road agencies.[21]

In the backdrop of the preceding discourse, the core proposition of this book is that the role of international institutions and their associated consultants shaped the outcome of transport planning in Pakistan. Sometimes their intervention limited the application of sustainable transport solutions consistent with Pakistan's need for social and physical development.

In summary, therefore, it appears doubtful if Pakistan would be able to institute sustainable transport policies, primarily because of its increasing reliance on international development institutions for funding and the continuing financial weakness of its provincial and city governments.

e) Research and Educational Capacity

Government organizations in Pakistan have played a key role in transport planning and creating the associated infrastructure. One way to assess the capacity of local transport institutions is to explore the quality and nature of research on transport, the education offered by local universities and the efforts made to raise awareness in the general public about issues of sustainable transport.

For conducting transport research, the NTRC was established within the federal Planning Commission in 1974. However, research in the NTRC is limited to the operational aspect of transport efficiency and to the structural and technological aspects of reducing capital requirement for building the infrastructure.[22] The Sustainable Development Policy Institute (SDPI)[23] was created in 1992 based on the recommendation made in the NCS[24] and given the charter to formulate sustainable development policies for Pakistan. Ever since its inception, programmes for sustainable transport have not received funds for research. The MTDF for 2005–10 proposed the creation of an Urban Policy Research Centre (UPRC) under the federal Planning Commission.[25] It was proposed that the UPRC would develop a national urban policy which would include a component for urban transport in the larger cities of Pakistan. The proposal to establish a research centre at the national level was based on the assumption that transport research should be the primary responsibility of the federal government, but that this function could be shared with the provinces and city governments.[26] In practice, research at the federal level cannot be accomplished without the technical and financial assistance of foreign consultants deployed by international development institutions. For example, planning standards for highways in Pakistan were prepared in 1962 by Howard, Needles and Nergendoff Consultants (American), and these continue to be adopted by all government organizations. Academics and researchers who were interviewed identified that:

> ...research is not in the list of government priorities. For example the National Transport Research Centre (NTRC) is only one of its kind in Pakistan. Initially NTRC was attached to the Planning Commission to provide input in the transport planning process, but the government has brought the centre to a lower level by placing it under the ministry of communication (Programme Officer, Pakistan Planning and Management Institute).

The above statement is based on the view that:

> ...more usually...there is no or very little role for research in professional institutes like the National Transport Research Centre (NTRC), Pakistan Institute of Development Economics (PIDE), National Institute of Transportation (NIT) in transport decision-making process in Pakistan (Programme Officer, Pakistan Planning and Management Institute).

As regards university education in transport planning and engineering, it was found that up to the late 1990s there was no undergraduate- or postgraduate-level specialized course on transport in Pakistan, even in the presence of well established engineering universities.[27] In early 2000, UET Lahore started postgraduate degree courses in transportation and environmental engineering and undergraduate courses in urban environment and transportation engineering. However, the main focus of the curricula is on road engineering, road construction technologies and traffic engineering. Few subjects relate to air pollution control and EIA for road development projects while subjects on public transport planning, non-motorized transport planning and sustainable transport planning and policy are not included.

Similarly, NED University of Engineering and Technology, Karachi also started a postgraduate degree course in transportation engineering and an undergraduate degree course in urban engineering in early 2000. The curricula of these degree courses cover subjects like transport planning (urban transportation planning, introduction to transportation planning and operations), public transport (public mass transportation, urban mass transportation), traffic and transportation engineering (traffic engineering and management, advanced traffic engineering & management and transportation engineering) and transport economics (transportation economics and transportation systems evaluation). Although the curricula in NED University are quite comprehensive and cover reasonable areas of transport, they are mainly designed towards teaching the operational requirements of different modes of transport. There is room to include subjects related to land use and transport planning, transport governance, urban transport policy analysis, non-motorized transport planning and approaches to sustainable transport planning.

The National Institute of Transportation (NIT), National University of Science and Technology (NUST),[28] University of Engineering and Technology, Taxila,[29] and the University of Karachi have started post-graduate programmes in transportation engineering and transport economics. However, these programmes are still in early stages of development.

Building awareness among the general public involves both imparting knowledge on key issues and offering possible solutions. All segments of society need to understand the impact of their transport behaviour so that they can make informed choices to reduce the adverse effects of transportation on the environment. During field work in 2004, it was found that issues related to the impact of transport on the environment were rarely investigated by government, private and community institutions. Usually local environmental protection departments were keener to raise the community's awareness of air and noise pollution. Few NGOs[30] worked for the promotion of road safety. Professional institutes like CILT Pakistan limited their activities to arranging one or two national-level workshops on transport, while some multinational private companies promote awareness of the benefits of using lead-free petrol. But there is total absence of an integrated effort amongst public, private, community and professional sectors to promote awareness of sustainable transport. In short, there is an acute paucity of research, mismatch of university curricula and lack of awareness in the community for promoting sustainable transport in Pakistan. The past neglect of transport research and education has resulted in a shortage of technical human resources that are essential if the capacity of government organizations is to be improved.

The analysis of the government's institutional capacity in transport policy and infrastructure delivery in Pakistan leads one to conclude that these institutions are hampered by limited financial autonomy and a serious lack of human resources which in turn permits international development institutions to exercise a strong influence in determining the course of transport development in the country. The analysis now turns to explaining the power of international and national transport institutions in the above context.

Explaining the Power of International and National Transport Institutions

The organizational chart described in Chapter 4 elaborates the transport decision-making structure in Pakistan. It shows that the primary responsibility to formulate the transport policy is vested in the federal ministries of communication, railways, ports and shipping, and defence. However, it is the ministry of finance and the ministry of planning and development—the two most powerful federal organizations—who assess the funding requirement for transport projects and accord approval for enlisting appropriate staff for development schemes.[31] Each ministry responsible for the transport sub-sector formulates policies in isolation, aiming to obtain the largest amount of funds for its own sub-sector.[32] In the current hierarchy, the provincial and local governments are responsible for executing policies of the federal government.[33] The NTRC transport policy[34] document states that provincial responsibilities are confined only to main roads (provincial highway department) and road transport (road transport corporation, provincial transport authorities). The local government responsibility is limited to local streets, and to issuing permits for non-motorized transport in the municipality. By virtue of controlling the finance, the federal government ensures that its transport priorities are properly implemented. For transport, the government provided 100 per cent of the allocated funds on the 'planning by directive' principle rather than 'financing as incentive'.[35] The 'planning by directive' principle means that there is no input from the provincial and local governments in the planning process and that their role is limited to merely implementing the policy formulated by the federal government. On the other hand the 'financing as incentive' principle implies providing transport solutions with incentives to attract investment. The NTRC transport policy document termed the practice of 'planning by directive' as a serious problem which results in a 'waste of time, effort and energy'.[36] No where in the world except perhaps in Pakistan would one find the practice of requiring repeated authorization from the ministry of finance to incur expenses previously approved and budgeted, sometimes even for expenses that are very minor.[37] The NTRC documents found that there is a lack of transparency in the transport decision-making process.[38] All PDWP, CDWP and ECNEC working papers are secret or confidential documents, and neither is the public consulted in the transport decision-making process. As a result, the public have had no say in transport investment to date. There is no statutory framework to get feedback from local and nationwide users on any development project.

In short, transport decision-making in Pakistan is a very centralized process dominated by the federal government which in addition also provides the transport infrastructure, operates transport and regulates the system. In other words, the federal government finances, executes and regulates the transport system, a situation that was recognized by all experts and academics interviewed for this study. As one expert stated:

> ...transport decision-making in Pakistan is in the hands of the federal government; while I strongly believe that urban transport policies should be addressed at the provincial and local levels (Deputy Chief, Physical Planning Section, Planning Commission).

However, the picture that emerges from the discussion of government institutional capacities is that transport institutions in Pakistan are technically and financially dependent on international development institutions. The volume of international aid and advice for transport development in Pakistan increased greatly between 1947 and 2005. Although it has been acknowledged several times that Pakistan's major problem has been its excessive dependence on foreign capital for financing its development programmes (see discussion in Chapter 3), no measures have been suggested (especially in transport) to reduce this dependence.[39] The relatively easy availability of external technical and financial assistance for transport development in Pakistan doesn't encourage the country from making its own decisions on future transport needs that suit local requirements. The dynamics of policy making have been greatly influenced by international development institutions, foreign experts and an intellectually dependent local bureaucracy. Two respondent interviewed explained this phenomenon as follows:

> ...Pakistan has a centralized transport policy making system where corporate groups and international institutions interact directly at the top bureaucracy level and make decisions. In this process, there is no public participation for transport decision-making (Associate Prof. of Environmental Planning, UET, Lahore).
>
> ...in Pakistan the bureaucracy holds the power to make policies with financial and technical assistance from international development institutions. Policies are formulated to suit the interest of local elites and Western donors. In the policy making process, politicians, civil society, academics and NGOs usually exercise little or no influence. In short, domestic elites and international funding agencies have made a joint coalition for policy making (Director, Institute of Policy Studies).

International institutions like the World Bank and its regional affiliates are now enormously influential, giving economic advice and formulating up policy. Ercelan's analysis demonstrates that international development institutions encouraged Pakistan to invest in 'hard infrastructure' rather than 'soft infrastructure' (institutional development).[40] This analysis fits the case of the transport sector very well, where international development institutions advised Pakistan to focus on building roads to achieve economic development, but failed to build institutional capacities. As one expert states:

> ...due to lack of financial resources, we are not able to set any priorities. We have to adopt whichever projects the donors offer because we need the money for development. Thus, most projects are 'donor driven' to fulfil donor priorities. Even now tree plantation in Pakistan is undertaken with the donor selecting the type of trees to be planted. For example IUCN is very active in Pakistan in the field of environment, but only for the promotion of a green environment as opposed to a brown environment. Their priorities can easily be gauged from our environmental policy documents (Deputy Chief, Physical Planning Section, Planning Commission).

These claims may be verified by exploring the various modes of transport that international development institutions have financed over time. Figure 5.1 shows the money spent by the IBRD and IDA on different transport sub-sectors from 1950 to 1993. The graph illustrates that financial assistance from the World Bank increased steadily in favour of road development from the 1950s to the mid-1960s and with the passage of time, roads clearly became the top priority for the IBRD and IDA programmes. Although in recent times the World Bank has started to place emphasis on non-motorized transport (transport expert, World Bank), the priority for economic development through road investment consumed the lion's share of its investment in developing countries; this then confirms one of the factors responsible for the road development paradigm in Pakistan.

It follows from the above discussion that Pakistan's transport development was made possible by the technical and financial assistance of international institutions. Foreign advice can be best utilized if indigenous institutions are strong whereas the most glaring weakness in Pakistan's transport policy was the failure to build institutional capacities. External economic assistance failed to strengthen government organizations at any level and neither did local academic institutions receive any substantial financial and institutional support to encourage research on developing a transport system in Pakistan that would be equitable, efficient and sustainable. Within the theoretical framework of this study, it has been argued that Pakistan's dependence on external technical and financial assistance allowed international institutions to strengthen their hold and influence on the country's transport policies. As a result of this reliance on international aid, Pakistan's transport policies became

FIGURE 5.1

IBRD and IDA Commitments to the Transport Sector (1950–93)

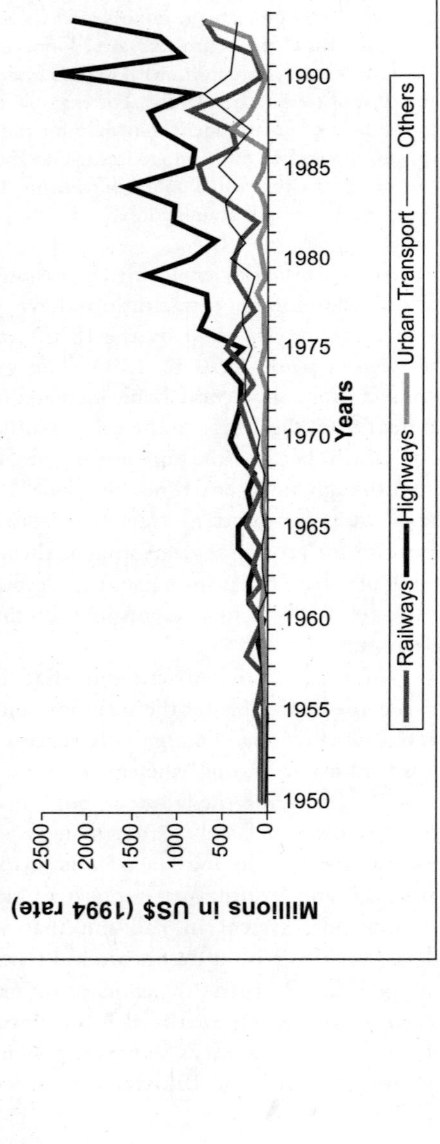

Source: World Bank 1994: Table A.3, 149

increasingly institutional path dependent because they remained consistent in subscribing to the rules, norms, expertise and organizational practices of international development institutions.

CRITERIA FOR DELIVERING TRANSPORT SOLUTIONS

Historically, there were three main criteria for investment in Pakistan's transport sector. The first criterion was the premise of 'grow, guess and gain'. Transport development in Pakistan started on the basis of a projection of its GDP growth (grow) and the assumption that this growth would generate a higher volume of traffic (guess) which would need additional transport infrastructure (gain). A variant of 'predict and provide',[41] the main assumption behind this criterion is the strong belief in the close relationship between the level of economic activity and the volume of transport.[42] Therefore, the first six Five-Year Plans forecast the volume of traffic demand on the basis of projected growth of GDP. However, these traffic forecasts varied with each Five-Year Plan.

In the presence of a comprehensive network of railways at the time of partition, the First Five-Year Plan (1955–60) proposed that the railway and road transport system operate side by side.[43] However, the Second Five-Year Plan (1960–65) suggested that road transport was particularly suited to the conditions and requirements of Pakistan. The suggestion was based on the argument that major population centres were located within short distances of one another and that the motor vehicle was more adaptable than the railways to the variations in traffic intensity resulting from this proximity, besides which it would permit a greater degree of speed and efficiency for freight and passenger transport.[44] In light of the 'grow, guess and gain' criterion and government policies towards road development, the traffic pattern favoured road transport over time. Due to the sharp increase of road traffic, the Five-Year Plans aimed at significant improvement and development of the road system to meet the demand and prevent creation of bottlenecks.[45] A similar approach was adopted in urban areas to provide relief from the existing shortage of roads as well as to cater for additional traffic requirements[46] justified by the urban transport planning process (UTPP).[47] In addition, government investment in transport was decided on the basis of the traffic forecast for each mode of transport because of which when the Fourth Five-Year Plan (1970–75) was launched, the plan suggested that the emphasis of transport investment should be shifted from railways to the road sector because the latter carried 50 per cent of the passenger and freight traffic.[48]

In the beginning it was assumed that long distance freight and passenger traffic would be carried by the existing railway. Every Five-Year Plan over the last fifty years projected the rail/road freight traffic ratio according to the expected GDP growth. However, it was found that the rail/road freight traffic ratio declined in favour of roads from 77:23 during the Second Five-Year Plan (1960–65) to 5:95 by the end of the Eighth Five-Year Plan (1993–98). The Planning Commission argued that the anticipated railway/road freight traffic ratio declined due to the lesser allocation of financial resources to railways.[49]

Before the beginning of the Fourth Five-Year Plan (1970–75), another criterion, 'decide and provide', was developed for road investment. This was based on the crude standard of 'road density' (road length in kilometres per square kilometres of area) irrespective of traffic volume. It was therefore decided that Pakistan needs 0.5 km of road length per sq km of area and that the shortfall would be made up in the long term.

The history of this criterion shows that the West Pakistan highway department engaged American consultants, Howard, Needles and Bergendoff, in 1962 to prepare the planning standards for developing and improving roads and highways. These standards were adopted by the West Pakistan highway department and the Planning Commission, and since then all highway planning has been done on the basis of these standards.[50] The Fourth Five-Year Plan (1970–75) stressed the need to adopt the common investment criteria in transport that was developed by the Planning Commission with the help of American consultants.[51] This criterion was formally adopted in the Fifth Five-Year Plan (1978–83).

The criteria were developed on the assumption that 0.5 km of road length per sq km of area is a generally accepted standard for developing countries with similar topography and level of economic development. In 1978 it was calculated that Pakistan had an average density of 0.16 km[52] of road length per sq km of land area, which was below the standard. On this crude formula it was decided to make up for this shortfall through a long term road construction programme stretching over many Plan periods.[53] The NHA defined the basic planning parameters for provision of roads in Pakistan in a presentation at the Pakistan Development Forum in 2005 as 'the balanced geographic distribution of investment according to road density criteria, with stress on uniform/standardized specifications of all highways irrespective of traffic volume'.[54] In Pakistan, where 80 per cent of the population lives along the N-5 (GT Road and Shahrah

Pakistan or North-South link), the criterion is not well supported by strong evidence of need.

In summary, the 'decide and provide' principle is presently the main criterion for road development and investment in Pakistan. It is now widely recognized that the practice of building highways cannot fulfil the objectives of sustainable transport. Roads alone cannot meet the increasing demand for transport, and even if they could, this is not seen as desirable for financial and environmental reasons.[55] Sustainable transportation options rely heavily on a plurality of methods in order to achieve multiple and integrated transport modes.[56] It is therefore critically important that future road development and investment in Pakistan should be based on a set of criteria that best suit the local environment and needs of the country.

MAIN STORYLINES IN ARGUMENTS FOR TRANSPORT SOLUTIONS

By reviewing the transport planning documents in Pakistan and interviewing key transport personnel who may or may not have been involved in the decision-making process, this study has identified storylines constructed around economic and environmental tales. These are narrated in the following text.

The Economic Tales

The storyline of transport development in Pakistan mainly runs around [economic] 'development'. For example, many experts believed that:

> ...Pakistan lacks development...there is scope for development...we are behind in development...development is necessary for Pakistan (Deputy Chief, Environment Section, Planning Commission).
>
> ...Transport is an important indicator of development (Presentation by the minister of environment, 6 October 2004).

To reinforce development storylines, the main supporting storyline was that 'economic growth needs transport infrastructure'. This was based on the assumption that the GDP/GNP growth would bring an increase in growth of transport which in turn would require transport infrastructure, e.g. 'as the GNP is being planned to grow at the rate of 6 to 7 per cent, the traffic is expected to increase at the rate of 10 to 12 per cent per annum'.[58] However, the transport development's main focus was on freight

TABLE 5.1

Comparison of Road Density in the World

Group	Country	Area (in sq. km.)	Paved Roads in km. (1997-2002)	Road Density (Roads length/ Area)	GDP per capita in US$ (2004)
High Income	Australia	7,713,000	308,409	0.04	31,690
Countries	Canada*	9,976,000	301,348	0.03	30,586
	Finland	338,000	50,336	0.15	35,562
	Norway	324,000	71,185	0.22	54,465
	Sweden	450,000	167,604	0.37	38,525
	USA	9,373,000	3,750,413	0.40	39.883
Middle	Brazil	8,512,000	94,871	0.01	3,284
Income	Malaysia	330,000	51,318	0.16	4,753
Countries	Philippines	300,000	19,201	0.06	1,036
	South Africa	1,221,000	57,677	0.05	4,675
	Thailand	513,000	56,541	0.11	2,539
	Turkey	779,000	147,441	0.19	4,221
Low Income	Bangladesh	144,000	19,711	0.14	406
Countries	China**	9,561,000	1,854,834	0.19	1,490
	Egypt	1,001,000	49,984	0.05	1,085
	India	3,288,000	1,889,627	0.58	640
	Nigeria	924,000	60,067	0.07	560
	Pakistan***	796,000	258,340	0.32	632

Source: Road data obtained from the World Bank (2005) and GDP data obtained from the UNDP (2006).

* Canada road data was obtained from Transport Canada website and represent 1995 figures.

** China road data represent 2004 figures.[57]

*** Pakistan road data represent 2006 figures (Govt. of Pakistan, Ministry of Finance 2006: 201).

movement, arguing that economic growth would bring a constant intra-urban flow of freight traffic. For example, 'proposed transport development is according to the requirement of agricultural and industrial programmes, urbanization and the need for balanced regional development'[59] or 'the inadequacy of transport is believed to be a serious obstacle to productivity and the growth of the economy'.[60] Transport documents stated that, 'in spite of spending of one-fifth of public sector funding, growth of the economy still remains handicapped by limitations in the transport systems'.[61] One researcher was critical of these assertions and argued that:

> ...Pakistan's development approach is rooted in the 'growth paradigm' to conform to the thinking of the international funding institutions. For example everything is measured from the criterion of GDP growth or mileage of roads in transport (Director, Institute of Policy Studies).

In contrast, all policy documents and publicity campaigns promoted the view that a strong economy will require an efficient and reliable transport system.

With the passage of time, the above storyline was turned around as: 'investment in transport infrastructure brings economic growth'. For example the Third Five-Year Plan stated that 'the pace of economic progress is largely determined by the adequacy and efficiency of the transport system' in Pakistan[62]...'transport sector provide infrastructural facilities to agricultural, industrial and other users of various services to accelerate the pace of economic development in the country'.[63] In short, the storyline was that the economy and transport development complement each other. Apart from these documents, government organizations promoted such perceptions in newspaper advertisements like:

> ...communication system can be considered a criterion for evaluating the development status of the country.
> ...a strong transportation and communication infrastructure has remained the cornerstone of all the modern day developed countries in pursuing their development objectives and in making their national policies.[64]

All documents and campaigns stress the linkage of efficient movement of transport with economic activity. However, the main emphasis then shifted quickly to road development. For example, in the past, the Third

Five-Year Plan clearly stated that in '[West] Pakistan, the road programme should receive greater priority.'[65] More recently:

> ...Pakistan needed a strong infrastructure particularly a modernly *(sic)* equipped and strong road network for realizing its developmental pursuits. Gradually the country has been developing highways, bypasses and motorways to meet the demands of time.[66]

Although there are some alternative perspectives like 'big budget projects are not enough to solve our country's transport problems',[67] such perspectives were not strong enough to change the overall mindset.

Road development as a 'vision of [economic] prosperity'[68] is well supported by all policy documents. According to the TSDI, 'construction of new roads should be decided strictly by economic prioritization'. It is also stated that 'economic considerations should govern all decisions on new roads'.[69] The document perceived that an increase in the road network would also provide opportunities to establish small and medium size businesses e.g. restaurants, repair shops, etc. The NTRC transport policy perceived economic sustainability as the prime factor in transport projects and stated that implementation of these projects within the designated time frame and allocated financial resources could achieve sustainability.[70] Here 'sustainability' simply refers to continuous growth of the economy. When the economic factor of the transport mode is a prime criterion then there is always a discussion on subsidy. Transport policy documents formulated in 1990s stressed that no subsidy would be provided for any mode of transport [keeping in mind the railways].[71] However, the same document describes its commitment to the road sector by stating that 'the freedom of the road haulage industry should be protected at all cost'.[72] Similarly it has been stated that 'the first priority should be achieving maximum benefits from investment already made', that is investment in roads. This again is in line with the prime perception of decision-makers that road development and the economy complement each other in contributing to the growth of the GDP.

This storyline is well assisted by international development institutions. This can be found mainly in the World Bank documents in which studies attempting to link aggregate infrastructure spending to growth of the GDP show very high returns in a time-series analysis.[73] It is evident that a strong association exists between the availability of infrastructure such as roads and per capita GDP. For example recent studies in the US on the impact of infrastructure investment on economic growth suggest that this investment provides startlingly high rates of return of up to 60 per cent.[75] The World Bank documents therefore declare infrastructure to be the 'wheels of economic activity'.[76] In contrast, Black was not convinced that any highway or transport facility would bring economic growth.[77] He cited Wilson's argument that transport investment could result in a negative, neutral or positive impact on economy.[78] In the negative case the investment might take finance away from other social sectors or might provide an opportunity to capture the local market. However, for the investment to be positive, it should address a hurdle in the economic activity resulting from a lack of transport infrastructure. In short, Black emphasized that one should not simply assume that investment in transport infrastructure would bring economic growth.[79]

The dominant 'road development will bring economic growth' storyline is supported by four other stories. The first is that 'motorization will bring more economic well being'. This is why policy documents make it a point to say that 'the per capita motorization is only 0.145 which is the lowest in the world'.[80] Irrespective of the authenticity of this figure, the same argument has been recognized by experts from international and corporate institutions:

> ...if Pakistan wants to become a developed country then it has to increase its motorization. (Transport expert, World Bank)
>
> ...Pakistan has room for car ownership because the country has a very low car ownership rate as compared to the developed countries [Director, Indus. (Toyota and Daihatsu) Motor Company]
>
> ...Pakistan has lots of potential for motorization. Our car ownership is very low, just 9 cars per 1000 people which shows that we are now in the pre-motorization stage in Pakistan. Therefore, our company has developed an 'Automobile Vision 2012', to establish a globally competitive automobile industry in Pakistan by the year 2012 by increasing the motorization level, by enhancing automobile production and increasing capacity of the industry. [Director, Indus (Toyota and Daihatsu) Motor Company]

These quotes suggest that there is a strong belief held by international institutions and the corporate sector in Pakistan to increase motorization to aid economic development. The result of these perceptions can easily be seen in the government's industrial policy paradigm and the growth of the automobile industry in the last decade. For example:

> ...the present government's policies and strategies are directed at promoting and enabling a competitive manufacturing base during the coming years.

> Confidence in these policies is reflected in the successful attraction of significant local and foreign investors and manufacturers, all striving to meet the transportation needs of the growing urban population and the rural areas.[81]
>
> ...auto industry in Pakistan is the highest growth sector industry; auto production grew 250 per cent during the past two years.[82]

These quotes reflect how industrial policies are designed to increase motorization in the country. Along similar lines, 'the government's recent leasing policy encourages people to buy a car, but there is no such policy to encourage the use of public transport.' (Chief, National Transport Research Centre)

To encourage use of motor cars, policies supported the storyline of eliminating 'congestion' and 'encroachment' on the roads. The focus on relieving congestion is supported by arguments that narrow and poor quality roads along with heterogeneous traffic reduce the speed of movement.[83] The congestion sub-storyline assumes significance when international development institutions support relieving congestion by calculating its cost to the economy. For example, according to the World Bank:

> In large and growing developing countries cities, such as Bangkok and Jakarta, vehicle congestion already gives rise to considerable environmental and economic costs. For Bangkok, it is estimated that if reduced traffic congestion permitted a 5 per cent increase in peak-hour vehicle speeds, the value of travel time saved would amount to more than $400 million a year.[84]

Therefore, 'expansion of transport infrastructure can reduce total population loads, as congestion falls, average vehicle speed rises, and routes are shortened'.[85] Transport policies have stressed that it was important to control overloading of freight transport for protecting roads, as good surfaces were required to ensure fast movement of traffic. In recent years, government schemes to offer soft loans for purchase of cars and to reduce import taxes are a series of steps aimed at promoting motorization in the country in the belief that this symbolizes a developed economy and is representative of a developed nation.

The second important supporting storyline—'poverty reduction through road development'—was introduced two decades earlier. The main concern from the early 1980s was to reduce poverty, and this was to be done by undertaking road development activities which was considered a means to tackle poverty directly.[86] This storyline is still in evidence. For example, the following quote from the prime minister's message clearly shows the government intention on poverty reduction through [road] transport development.

> ...development of the transport sector is an important element of the government's poverty reduction strategy. Poverty remains a serious concern in Pakistan. If the transport sector is to prosper and to be effective in the fight against poverty, transport must be safe for all, be affordable, provide efficient mobility, offer equitable services and opportunities and focus on the poor.[87]

In light of this vision, the government not only promotes road development but also encourages the automobile industry in order to provide opportunities for employment and reduce poverty. For example:

> ...automobile industry in Pakistan has attracted Rs52 billion investments in 2004 which will increase up to 93 billion in 2006. Moreover, the automobile manufacturing sector is providing 116,500 jobs in 2004 which will increase up to 190,000 in 2006. Auto industry tax has increased from Rs6 billion in 1997 to Rs24 billion in 2004. [Director, Indus (Toyota and Daihatsu) Motor Company]
>
> ...auto industry in Pakistan employed directly 160,000 persons and provides indirect livelihood to 1,000,000, contributed Rs31 billion to government revenue or 8 per cent of total government receipts, and invested Rs52 billion.[88]

Pakistan's policy for poverty reduction is in line with the World Bank criteria for funding transport development. The World Bank strategy for poverty reduction in developing countries is based on the three pillars of the Bank's urban transport strategy: (1) improving the operational efficiency of transport (2) better focusing of interventions to assist the poor, and (3) policy and institutional reforms in terms of infrastructure pricing, service provision and pricing, urban transport system financing, strengthened institutions and sound political framework (transport expert, World Bank). The World Bank has also supported the 'poverty reduction' storyline in their World Development Reports: 'the construction and maintenance of some infrastructure especially roads can contribute to poverty reduction by providing direct employment'.[89] It is very evident that road development storylines are embellished as nation building schemes, as for instance, 'nations do not build the roads; it is roads that build the nations'.[90]

The Environmental Tales

In the decade of the 1990s, social and environmental concerns of transport received greater attention. These concerns were expressed in the TSDI transport policy objectives in 2001: 'to move people and goods efficiently, economically and safely in an environmentally sustainable framework'.[91] It went on to say that 'transport policy must set realistic socio-economic development goals achievable within the available resource and other constraints...must be comprehensive and fully integrated, must be responsive to the needs of the users in general and of the most vulnerable segment of the society in particular'.[92] These objectives are strengthened in the NTRC 2001 transport policy document which stated its vision as:

> A system that provides efficient, safe, reliable, equitable and environmentally friendly access and mobility for people and goods thereby supporting the governmental goal of increasing public welfare through economic growth, human development and poverty reduction.[93]

The document further stated that 'investment in transport infrastructure should provide economic priorities, social equity and environmental sustainability'.[94] Hence the strategy was developed on the basis of 'moving people not vehicles'.[95] The CILT 1998 transport policy also gave the environment a high priority.[96]

However, environmental concerns revert to 'regulatory measures, technical fixes and road safety' storylines. In the beginning the main stress was to enact various laws and formulate programmes to control transport emissions only. For example, the section 15 of the Pakistan Environmental Protection Act, 1997 set priorities of Pak-EPA for the transport sector. It legislates the formulation of rules and regulations to address pollution issues arising (only) from transport. However, detailed rules and regulations have not yet been formulated. Many experts believe that 'by enforcing Pakistan's Environmental Protection Act, 1997, we can achieve our priorities for sustainable transport' (Director, Pakistan Environmental Protection Agency) and we can 'achieve clean air by enforcing rules and regulations' (Deputy Chief, Environment Section, Planning Commission).

The 'technical fixes' storyline runs around the promotion of CNG cars, lead-free petrol, catalytic converters, tune-up stations or car inspection centres, induction of new cars, modernization of the auto industry, improved engine performance and instituting air quality measurements.

But they do not consider the full ecological cost of roads and road traffic. These storylines are verified in the President of Pakistan's message:

> Pakistan automobile sector today also faces critical challenges, which include support to indigenization, human resource development, exhaust and emission controls, improved engine technologies, clear fuels, systematic inspections, maintenance and certification systems, phasing out of old vehicles, introduction of integrated transport management systems and policies, environment friendly manufacturing and operating systems and provision of equitable, accessible, affordable transport to the rural and the urban areas.[97]

These technical fix storylines are supported by international development institutions like the JICA which provides technical and financial assistance to control air and noise pollution from road transport (environment and transport expert, JICA). In addition, making new roads and related infrastructure to relieve congestion are also supported on environmental grounds by arguing that free movement of traffic would cause lesser pollution.

The 'road safety' storyline is based on the perception that road accidents cause a major economic loss, with CILT estimating the loss at Rs45 billion per year which does not take into account the social and emotional loss to the families of road accident victims.[98] Road users' behaviour has been cited as the 'root cause' of road traffic crashes in Pakistan. Driver training, use of seat belt and helmet, provision of quality roads, improvement of geometric design, and vehicle fitness are some of the suggested measures for improving road safety. Some enforcement measures like implementation of the National Highway Safety Ordinance 1997 and introduction of industrial reforms involving modernization of the auto industry are also proposed.

In short, environmental storylines have been deployed to build more roads and promote motorization. For example, senior experts at the national level stated:

> ...we cannot stop road construction in the name of environment. However, we are trying that environment will be incorporated into road development (Deputy Chief, Environment Section, Planning Commission).
>
> ...Pak-EPA has no plan to discourage cars because we have not reached the level of car ownership as in Los Angeles. We have 160 million people and only one million vehicles; we are far behind developed countries. However we want clean fuel, and standard vehicle condition for clean air in Pakistani cities. Our problem is emission and enforcement only (Director, Pakistan Environmental Protection Agency).

These storylines are well supported by the World Bank stating that, 'well-designed and managed infrastructure can promote environmental sustainability'.[99]

This discussion may be surmised with the conclusion that the economic and environmental storylines are used to justify an increase in building new roads. For example, Husain sums this up by stating that infrastructure can contribute to economic growth, poverty alleviation and environmental sustainability but only if an adequate supply of transport infrastructure has been provided.[100] He supported this by arguing that a 1 per cent increase in the stock of infrastructure is associated with 1 per cent increase in the GDP across all countries.[101] Husain's claim has been developed from the World Bank view on infrastructure.[102]

NOTES

1. Waterston 1963: 34.
2. Husain 1999: 258.
3. Grindle 1996: 7.
4. Jenkins and Smith 2001: 485–507.
5. Govt. of Pakistan, EUAD and IUCN 1992: 92.
6. Govt. of Pakistan, SMEDA 1999.
7. Meakin 1998.
8. Husain 1999: 259.
9. Ibid., 375.
10. LaPorte 1984: 242.
11. Waterston 1963: 34.
12. Ibid., 36.
13. Fatima 2003.
14. Govt. of Pakistan, NTRC 1991: 45.
15. Govt. of Pakistan, Planning Commission 2000: 5.
16. Waterston 1963: 27.
17. Ibid., 137.
18. Husain 1984: 105.
19. Ibid., 105.
20. Burki 1992: 129.
21. ADB 2005: 28–30.
22. Govt. of Pakistan, NTRC 1991: 41.
23. www.sdpi.org
24. Pakistan Agenda 21.
25. Govt. of Pakistan, Planning Commission 2005: Urban Development and Housing, 9.
26. Govt. of Pakistan, NTRC 1992.
27. In Lahore, the University of Engineering and Technology (known as Maclagan Engineering College) was established in 1923. Later, the college was affiliated with the University of the Punjab for award of a bachelor's degree in civil, electrical and

mechanical engineering (http://www.uet.edu.pk/). In Karachi, the NED University of Engineering and Technology (known as Prince of Wales Engineering College) was set up in 1922. Later, the college was affiliated with the University of Bombay for award of a bachelor's degree in civil engineering (http://www.neduet.edu.pk/)
28. http://www.nust.edu.pk/R&D.htm
29. http://www.uettaxila.edu.pk/civilPGprog.htm
30. http://www.pakroadusers.com/
31. Govt. of Pakistan, NTRC 1991: 47.
32. Ibid., 48–49.
33. Ibid., 52.
34. Ibid., 52.
35. Govt. of Pakistan, NTRC 1991: 48–49.
36. Ibid., 1991.
37. Ibid., 48.
38. Govt. of Pakistan, NTRC 2001: 17.
39. Govt. of Pakistan, Planning Commission 2001: 18.
40. Ercelan 2004: 3–4.
41. Owens 1995.
42. Govt. of Pakistan, Planning Commission 1960: 277.
43. Govt. of Pakistan, National Planning Board 1957: 485.
44. Govt. of Pakistan, Planning Commission 1960: 303.
45. Govt. of Pakistan, Planning Commission 1970; 1978; 1983.
46. Govt. of Pakistan, Planning Commission 1978: 10.
47. Dimitriou 1992.
48. Govt. of Pakistan, Planning Commission 1970: 225–229.
49. The Planning Commission 2000: 7.
50. Govt. of Pakistan, NTRC 1992: 12.
51. Govt. of Pakistan, Planning Commission 1970: 231.
52. These figures are different than Table 5.1. There is a chance that unpaved road was also included in this figure.
53. Govt. of Pakistan, Planning Commission 1983: 265.
54. Govt. of Pakistan, NHA 2005, Pakistan Development Forum 2005 presentation.
55. Banister 2002.
56. Newman and Kenworthy 1999.
57. http://www.adb.org/Documents/TARs/PRC/tar-prc-38174.pdf#search='TAR%20PRC%2038174%20TECHNICAL%20ASSISTANCE%20TO%20THE%20PRC'
58. Govt. of Pakistan, Planning Commission 1970: 239.
59. Govt. of Pakistan, Planning Commission 1965: 45.
60. Ibid., 137.
61. Govt. of Pakistan, Planning Commission 1970: 223.
62. Govt. of Pakistan, Planning Commission 1965: 137.
63. Govt. of Pakistan, Planning Commission 1978: 38.
64. *Dawn*, 21 September 2004—NHA Advertisement.
65. Govt. of Pakistan, Planning Commission 1965: 45.
66. *Dawn*, 21 September 2004—NHA Advertisement.
67. *The Nation*, 6 October 2004—Chairman Senate's message to the 'National Conference on Sustainable Transport—The way forward'.
68. Media coverage for inauguration of road projects, *The Nation*, 21 September 2004.
69. TSDI 2001.
70. Govt. of Pakistan, NTRC Transport Policy 2001.

71. Govt. of Pakistan, NTRC 1991.
72. Ibid., 12.
73. Govt. of Pakistan, NTRC 1992.
74. World Bank 1994: 14.
75. Ibid., 14–15.
76. Ibid., 14.
77. Black 2001: 3.
78. Wilson 1966.
79. Black 2001.
80. Govt. of Pakistan, NTRC 1992: 28.
81. *The Nation*, 6 October 2004—President's message to the 'National Conference on Sustainable Transport—The way forward'.
82. *Daily Jang*, 18 June 2004—Pakistan Association of Automotive Parts and Accessories Manufacturers.
83. Govt. of Pakistan, Planning Commission 1965: 143; Husain 1999: 245.
84. World Bank 1994: 21.
85. Ibid., 22.
86. Govt. of Pakistan, Planning Commission 1983: 10–12.
87. *The Nation*, 6 October 2004—Prime Minister of Pakistan's message to National Conference on Sustainable Transport—The way forward.
88. *Daily Jang*, 18 June 2004—Pakistan Association of Automotive Parts and Accessories Manufacturers.
89. World Bank 1994: 20.
90. Govt. of Pakistan, NHA 2005, Pakistan Development Forum 2005 presentation.
91. TSDI 2001: 3.
92. Ibid., 3.
93. Govt. of Pakistan, NTRC 2001: 3.
94. Ibid., 5.
95. Ibid.
96. CILT, 1998: 6.
97. *The Nation*, 6 October 2004—President of Pakistan's message to the 'National Conference on Sustainable Transport—The way forward'.
98. CILT 1998: 11.
99. World Bank 1994: 20.
100. Husain 1999: 243–4.
101. Ibid., 243.
102. Husain 1999; World Bank 1994: 15.

6

Urban Transport in Pakistan: A Case Study of Lahore

The urban population of Pakistan grew from 5.9 million in 1951 to 43 million in 1998,[1] which was 32.5 per cent of the country's total population (135.8 million) in that year. This means that every third person then lived in a city or town in Pakistan. By the year 2020, the urban population is expected to be over 50 per cent of Pakistan's total population.[2] Urban settlements occupy less than 0.8 per cent of Pakistan's land area which houses 32.5 per cent of its population and contributes almost 48 per cent of its GNP.[3] In 1998, eight cities of the country had populations of over one million with Karachi having the largest number (9.2 million) followed by Lahore (5.4 million). The dynamics of these cities is also evident from their growth rate; Lahore for example grew by an average of 3.3 per cent between 1981 and 1998.[4] Lahore has been selected for this case study because it is widely regarded as the city where new and forward-looking policies are implemented to validate their efficacy in the Pakistani context. It might be expected, therefore, that these policies would manifest themselves in Lahore's transport policies. However, the hypothesis that the transport policy at the national level is path dependent, would be diffused in Lahore.

The chapter first looks at Lahore's general characteristics, and its urban transport issues and challenges to provide the reader with a context with which to determine the applicability of the conclusions to other cities. General characteristics include population, area and land use in both the historic and present context. Transport issues such as the contradiction between travel behaviour and transport priorities, insufficient public transport, rapid growth in private vehicle ownership and the deteriorating environment will be described to set a context for the urban transport situation in Lahore. The following section examines urban transport policies in Lahore during the period from 'British' Lahore to 2005. The chapter then examines the current transport policy making process in

Lahore in greater depth. The focus is to detail the respective areas of responsibilities amongst the various provincial and local institutions involved in transport policy-making for Lahore. The capacity of highway/roads, public transport and the planning institutions will be explored to determine the role of local transport institutions in providing transport and in devising the transport policy for Lahore in association with federal and international institutions. The chapter will set the context for discussion on a typical road proposal—the Ring Road in Lahore (in Chapter 7)—to examine whether there is any path dependence manifest in the process.

INTRODUCING THE CITY OF LAHORE

Lahore, the provincial capital of Punjab, is not only the second largest city of Pakistan but also its cultural centre. It is situated in the north-eastern part of Pakistan on the left bank of the River Ravi and is only 25 km from the international border with India as shown in Figure 6.1. It has grown along the historical route of the Grand Trunk (GT) Road linking Central Asia with the subcontinent. The main railway line connects Lahore to most cities along the northern and southern routes and to India in the east.

EVOLUTION OF LAHORE

The evolution of the Lahore metropolis dates back to the first millennium. According to Hindu tradition, Lahore was founded by Loh, son of Rama. For the first time in recorded history, Lahore appeared in the memoirs of the famous Chinese traveller, Hiuen Tsiang, who mentioned it as a large Brahmin city visited by him in AD 630 while on his way to Jullundur (now in India).

The first Muslim rule was established in Lahore in AD 988 when the ruler of the kingdom of Ghazna (now in Afghanistan) invaded the state of Punjab. The golden period of Lahore started with the establishment of Mughal rule in India in AD 1524. During the rule of the Mughal dynasty, Lahore became a great centre of learning, knowledge, architecture, arts and crafts. Lahore became a 'city of gardens' after the Mughals constructed many gardens within and around the grand wall of Lahore. Amongst those, the Shalimar Garden (1642) is the most famous. The Lahore Fort (1566), Wazir Khan's Mosque (1634), Mughal Emperor Jahangir's Mausoleum (1637) and Badshahi Mosque (1673) are unique in architectural design and construction and have now become world

FIGURE 6.1
Location of Lahore

Source: adapted from http://www.un.org/Depts/Cartographic/map/profile/seasia.pdf

heritage sites. There is no doubt that the Mughal dynasty made Lahore into a centre of vital importance.

British occupation of north-western India in 1846 brought modern civilization into the historical and cultural city of Lahore. Lahore was connected to other parts of India by building a railway network. The magnificent building of Lahore railway station was completed in 1861, marking an important milestone in the development of modern transport in the subcontinent. The British established prestigious universities, medical, engineering, and commerce colleges and research institutes in Lahore. The more significant of these are the Government College (1861), the Punjab University (1868, even though the University Act was formally passed in 1882), Kinnard College for Girls (1879), Lahore Medical College or King Edward's Medical College (1883), Punjab Chief's College (1886), Forman Christian College (1883), Lahore College for Women (1922), Maclagon Engineering College (1923), Irrigation Research Institute (1925), Hailey College of Commerce (1927) and de' Montmorency College of Dentistry (1934).

Lahore got its modern municipal, social and judicial system with the construction of the Municipal Committee (1862), Montgomery Hall (1866), Mayo Hospital (1870), Lahore Water Works (1882), Punjab Public Library (1885), Chief Courts or High Courts (1889) and finally the Lahore Improvement Trust in 1936. The British enriched the list of gardens in Lahore by adding the Lawrence Garden (1862) and the Lahore Zoo (1872) to the pre-existing list of Mughal gardens. In addition, a modern print and telecommunication media were established in Dyeing and Calico Printing and the Broadcasting Housing in 1923 and 1934 respectively. The modern road network named after British deputy commissioners of Lahore (such as Mall Road, Egerton Road, Davis Road, Nicholson Road, Circular Road and McLeod Road) not only connected these major civic, educational and institutional buildings but also provided people with space for social activities under the shade of rows of trees planted along these roads. In short, at the end of British rule in the Indian subcontinent in 1947, Lahore became the centre of historical, cultural, social, educational and institutional activities.[5]

The partition of the Indian subcontinent brought a major upheaval in the region, and Lahore witnessed lasting changes in its socio-economic and physical sectors. Lahore City [old Municipal Corporation Lahore (MCL)] covers an area of 250 sq km and its population grew from 0.86 million in 1951 to 5.4 million in 1998 at the average annual compound growth rate of 4 per cent.[6] In 2006, it was estimated that 6.8 million people lived in Lahore.[7]

Land Use and Urban Structure

The layout of Lahore city represents three main zones. These zones stand out distinctly in terms of the generalized built environment and population density. These may be classified as central, intermediate and peripheral areas. The central and inner zones generally house the poor and middle class people, but these are also the main hub for commerce and business, employment, light industry, civic and administrative activities, and have a few scattered recreational facilities. The gross residential density of the inner zone is 250 persons per hectare or above.[8] The intermediate area (largely planned) is an uncontrolled mixture of housing (for middle and upper income groups) and related activities and services (i.e., education, health, recreation, utilities, etc.) but there remains a considerable dependence on the central zone for specialized commercial, business and employment activities. As a result, there are problems of commuting between these zones. The intermediate zone comprises an average density of 151 to 250 persons per hectare.[9] The outer area represents the typical characteristics of an urban sprawl where the city has grown through a combination of leaping out and filling in process. Planned and officially sponsored development extends to the urban boundaries. Private housing schemes are still being developed extensively in the outer zone where the average population density is less than 150 persons per hectare.[10] However, one study says that 80 per cent of the people still live within a 7 km radius of the city centre.[11] This area covers the inner, high density zone and most of the intermediate density zone.

According to the land-use survey 2002, the Lahore Metropolitan Area (LMA) is spread over 230,670 hectares, although most of it (76.2 per cent) is used for agriculture. The boundary of the erstwhile MCL covers an area spread over 250 sq km, of which 51 per cent is for residential use followed by 15.9 per cent for agriculture.

URBAN TRANSPORT IN LAHORE: ISSUES AND CHALLENGES

Lahore is the hub of commercial, social, cultural and political activities in the Punjab and contributes significantly to the national economy. However, these activities are under threat due to the growing problems of urban transport characterized by the following issues: contradiction between travel behaviour and transport policies, insufficient public transport, rapid growth in vehicle ownership and a deteriorating environment.

The Contradiction Between Travel Behaviour and Transport Priorities

The urban transport needs of a city are cyclic in nature and largely depend on the travel characteristics of its citizens. According to the TEPA and JICA studies in 1992, the population of Lahore generated about 12.4 millions trips a day of which 60 per cent were non-motorized. From amongst these non-motorized trips the largest proportion was that of walking trips (51 per cent) followed by bicycles at 8 per cent. In terms of motorized trips, cars and motorcycles contributed 12 per cent for each mode followed by 14 per cent for public transport. TEPA and JICA (1992) is the only comprehensive study conducted in Lahore which has calculated the number of trips for each mode. It has been fifteen years since this study was carried out, so I have assumed that the proportion of non-motorized trips in Lahore has decreased. But I consider that half the total trips are yet made by non-motorized transport.

Despite the fact that a large percentage of trips are made by foot, there has been no serious effort for pedestrian planning in Lahore. Provision of narrow footpaths along major roads is considered enough. It was observed that most of the existing footpaths have either been encroached upon or are ill maintained. In the entire city of Lahore, only one pedestrian crossing underpass and a few overhead bridges have been provided. But even these are not used because of their inappropriate location and poor maintenance of median barriers. Pedestrians not only have to suffer air and noise pollution, but also must put their lives at risk as most victims of road accidents are pedestrians.

Insufficient Public Transport

Public transport operations in Lahore are currently being delivered entirely through the private sector. Wagons (18-seater, small motorized vehicles) and buses are the main modes of public transport in Lahore. Transport and environmental stakeholders in Lahore (wagon union, rickshaw union and environmental groups) identified lack of public transport as one of the most important transport issues in Lahore. It was observed that public transport does not have sufficient capacity to carry large numbers of passengers, especially during peak hours and on busy routes. At peak times travel is extremely uncomfortable and unsafe because of overloading. The low quality of wagon service and old buses deters many people from using public transport, particularly women, elderly people and children. In the absence of proper governmental regulations, there is no control on the operation timetable, usage and condition of privately owned and operated vehicles. On the other hand, representatives of bus companies that were interviewed blamed the government and its policies for this situation. They argued that in the absence of public-transport-friendly policies, and institutional and infrastructure support, they are unable to provide a better public transport service. Overall, the ratio of available seat capacity on public transport to population in Lahore is very low which underlines the deficiency in the capacity of public transport.

The Punjab government initiated a franchised bus scheme in 1998. On some routes franchise operators enjoy a monopoly and have replaced wagon operations. In turn they guarantee a minimum quality of service albeit on a higher fare that is decided by mutual agreement. Currently there are twelve franchised companies operating nearly 700 buses in Lahore. Daewoo, the only international franchised bus company operating in Pakistan, provides air-conditioned service on three different routes. However, this scheme has not fully succeeded in attracting large scale investment from the private sector and the international development institutions. I found that because of the lack of institutional capacity in the government, it has not been able to play an effective role in attracting private investment and resolving conflicts between franchised bus and wagon operations. This is evident in the case of the public transport franchising operations in the twin cities of Rawalpindi/Islamabad which were eventually terminated. In Lahore too there are many conflicts amongst private bus-franchised companies and the government. For example, in the interviews conducted with Messrs Daewoo and the New Khan Metro both correspondents demanded government subsidy for providing low fares to students and disadvantaged people.

There have been several attempts in the past to introduce rail-based public transport in Lahore. Comprehensive studies conducted by TEPA and JICA in the early 1990s proposed a Light Rail Transit (LRT) system besides also suggesting the use of Pakistan Railway as an urban transport mode. But no government took serious steps to implement their proposal because of its high cost and opposition from influential private operators of public transport. In 2005, the Lahore Rapid Mass Transit System (LRMTS), a rail-based 4-line network was proposed on the assumption that people were able to pay for a better service on account of their rising income, something that had already been established by the success of the air-conditioned franchised buses introduced earlier. However, no evidence is provided in support of this argument. In short, public transport in Lahore remains in the hands of the private sector with little regard to the quality its service. On the other hand, public sector finance has been entirely devoted to roads.

Lahore is renowned for its approximately 13,000 noisy and smoky rickshaws.[12] They have the ability to enter congested and narrow streets because of their small width. The environmental protection department of the Punjab government has initiated several campaigns over many years against the air and noise pollution caused by rickshaws, but in the absence of an alternative public transport that is economical, it has not been entirely successful in phasing them out.

Rapid Growth in Private Vehicle Ownership

Heavy motorized traffic has become a part of everyday life in Lahore because of the sharp growth in the number of private vehicles over the last few decades. For example, there were only 39,205 licensed and 'on road' vehicles in the Lahore district in 1974.[13] These numbers rose to 652,082 in 1996[14] and reached an approximate figure of 890,096 in 2006.[15] Vehicles in Lahore mainly comprised of motorcycles (53 per cent) and cars (29 per cent).[16] In the past twenty years, the overall number of motorcycles has been higher than cars, but in the same period, against a three-fold increase in the number of cars, the number of motorcycles has increased two-folds. These figures represent the position in the Lahore district, but one can assume that traffic growth in the larger Lahore Metropolitan Area would be similar to that represented by the above figures.

There are many reasons for the sharp growth in the number of private vehicles in Lahore. Perhaps the most important amongst these are rapid population growth, housing development in the periphery of the city, unavailability of alternative transport, favourable policies for ownership of private vehicles and heavy investment in roads. Several new roads have been made in the outlying urban areas of Lahore which have made it possible for affluent people to live in the suburbs and travel long distances to get to the city. Similarly, several old roads have been widened in existing urban areas to reduce congestion, and the process continues. But this road-building exercise has been carried out at the cost of the majority because it serves only a small percentage of people that live on the outskirts of the city and can travel by car. Additionally, the government has provided tax incentives to financial institutions for car leasing schemes as part of its overall economic strategy. This policy too has resulted in a sharp increase of car ownership in the city.

The steep growth in the number of private vehicles has caused severe road congestion in Lahore. Almost all stakeholders who were interviewed identified 'congestion' as the major problem of transport in Lahore. For example, in the interviews conducted with Messrs Daewoo and New Khan Metro both correspondents supported building underpasses, flyovers and ring roads in order to relieve congestion. However, correspondents from the wagon union and the rickshaw union made congestion as a case for the use of wagons and rickshaws as efficient modes of public transport in the narrow and congested roads of Lahore. To summarize, the rapid growth in the number of vehicles, the absence of reliable public transport, illegal parking and poor traffic behaviour have in combination exacerbated the problem of congestion.

The Deteriorating Environment

The rapid increase in population, the absence of decent public transport and the high growth in the number of private vehicles have had a negative impact on the air quality of Lahore. According to the latest data compiled by the environment protection department (EPD) in association with JICA, carbon-monoxide (CO) and suspended particulates matter (SPM) concentration in central parts of the city was found to be ten times higher than WHO standards.[17] Most of the particulate matter can be attributed to the result of emissions from vehicles, re-suspension of dust raised by vehicles travelling on unpaved and under construction roads and idle burning of fuel.

Motor vehicles account for 90 per cent of the total emission of hydrocarbons, aldehydes and carbon monoxide in cities, and for three-fourths of all sulphur dioxide (SO_2) and nitrogen oxide (NO_2). Suspended particles in the air cause lung irritation, sinus problems, bronchitis and asthma, whereas lead emitted by gasoline vehicles causes liver dysfunction and impairment of mental health. Pakistan Strategic Country Environmental Assessment Report prepared by the World Bank[18] estimated that there are 22,000 premature deaths amongst adults and 700 deaths amongst children under 5 years annually due to air pollution. Although these figures are for the entire country they allow us to estimate the scale of the local air pollution problem.

In addition, data shows that a noise level of 89 to 90 decibels is commonly experienced on all major roads of Lahore, which exceeds the maximum advisable limit of 85 decibels.[19] According to the EPD, several transport related policies and projects were initiated after the creation of the department but these were introduced piecemeal and without taking into consideration the transport and land-use policies. As a result these projects failed to improve the environment of the city.

The continuous improvement and widening of roads in Lahore led to a corresponding growth in the rate of road accident fatalities (11.4 per cent) and injuries (10 per cent) between 1990 and 1996, with most victims being pedestrians or cyclists.[20] This is an alarming situation which especially discourages pedestrian and non-motorized traffic to play a role in the development of city transport.

TRANPORT POLICIES OF LAHORE IN HISTORICAL CONTEXT

Lahore in British India

Transport planning in Lahore started with the construction of the Mall Road in 1851.[21] When Lahore was made the capital of the British Punjab in 1858, government institutions relocated their offices along the Mall Road[22] which provided the link between the cantonment of Mian Mir (now called Lahore Cantonment) on one end to major civic and government buildings on the other. Many writers extolled the Mall Road in Lahore as one of the finest public roads in British India.[23] It not only gave access to all major government offices but also provided space for social activities and entertainment to people of all races and class.[24] According to Qadir, the Mall Road of the past was lined with many trees and was the venue of numerous social activities, adding to the overall beauty of Lahore.[25]

In 1870, the first topographical survey was carried out for developing the main roads of Lahore.[26] The Punjab Municipal Act of 1884 provided the first legal framework for municipalities to oversee building and street plans in Lahore. A street-widening project in the city was started in 1907 with several typical cross-sections of major and minor roads drawn up from different locations across Lahore.[27]

Patrick Geddes, a famous Scottish town planner of the early twentieth century, worked as a consultant, and prepared *Town Planning in Lahore—A Master Plan* published in 1917. From the start Geddes paid high regard to the city's potential, saying:

> What a gorgeous old city is Lahore, far finer than anything we saw together, for though of course without the sublime slope and river and spirit of Benares, it is the heaped up, crowded, lofty massing of centuries of the Mohammedan empire's wealth and art.[28]

Geddes adopted an approach of 'conservative surgery' by combining physical and social planning. However, he was afraid that if planners adopted foreign solutions and methods for Lahore then there was a chance that it might lose its unique characteristics. Around the time of Patrick Geddes' visit, Ebenezer Howard's idea of *Garden Cities of Tomorrow* (1898) had been circulating in Lahore. After Geddes' visit, Lahore's Model Town was planned which was almost a replica of the transport and land-use planning ideas of Howard's *Garden Cities*.[29] Later, the Lahore Improvement Trust was created in 1919 to oversee all development projects in the city.

Master Plan for Greater Lahore (1965–80)

Post-independence development of Lahore took place towards the south and south-west along four main arterial roads. The 'Master Plan for Greater Lahore' was prepared by the housing and physical planning department of the Punjab government in 1965. The Plan proposed urban Lahore to be developed as a metropolitan core with self-sustaining industrial towns in its periphery. In order to segregate urban Lahore from its surrounding towns, a 24 km wide green belt was proposed. For the transport network, the Plan proposed the construction of a 'Ring Road', a new road-bridge over the River Ravi, upgrading of major roads, grade-separated junctions over railway crossings and provision of underground and multi-storeyed parking lots in the central area. For commuter traffic, it was proposed that circular railway be introduced to provide the city with a mass transit system. For non-motorized traffic, the Plan proposed the gradual phasing out of animal-drawn carriages, yet encouraged the use of bicycles by proposing the provision of cycle tracks along major roads. The Plan also proposed setting up of a traffic planning department within the Lahore Improvement Trust. Most road network proposals in the Plan have been implemented, except for the Ring Road. However, recommendations for introducing circular railway as a mass transit system, and segregation of cyclists from the main traffic failed to draw the attention of the decision-makers.

Lahore Urban Development and Traffic Study (LUDTS) (1980–2000)

The Lahore Urban Development and Traffic Study (LUDTS) was prepared by foreign consultants with financial help from the World Bank in 1980. The study identified issues related to traffic characteristics, urban road infrastructure, institutions, and public transport operation and management. LUDTS proposed that a Traffic Engineering and Transport Planning Unit (TEPU), now Traffic Engineering and Transport Planning Agency (TEPA) in LDA, be established to deal with the city's traffic engineering and management issues. The other main proposal of LUDTS was to construct the 'Southern Bypass', renamed the 'Lahore Ring Road'. TEPA subsequently prepared and executed a number of motorized-traffic engineering proposals along with preparing feasibility studies for the 'Lahore Ring Road'. However, it failed to address non-motorized traffic such as pedestrians and horse-driven *tongas* in an integrated master plan. TEPA focused more on 'traffic engineering' rather than 'transport planning' for which this agency was originally created.

Model Urban Transport System in Lahore 1980

The 'Model Urban Transport System' in Lahore was completed in 1980 with the financial and technical grant of the Volvo International Development Corporation.[30] This study provided a trip generation model as a part of the exercise for estimating travel demand. The study identified various issues for providing an efficient bus-based public transport network in Lahore, and recommended a mix of public and private bus systems. As a result of this study, 350 Volvo buses were gifted by the Swedish government to Lahore. These buses were added to the fleet of Punjab Road Transport Corporation (PRTC), a public sector corporation responsible for bus-based public transport in the city. However, after operating for a few years, the bus system in Lahore collapsed because of organizational inefficiency of the government-sponsored PRTC. The PRTC was disbanded in 1998.

Punjab Urban Development Project (PUDP) 1988–97

The Punjab Urban Development Project (PUDP) was executed in various cities of the Punjab between 1988 and 1997. This project was financed by the World Bank. The major thrust of this project was directed towards improving the road geometry in Lahore. The project was executed through

Ground breaking ceremony of Lahore Ring Road. President General Pervez Musharraf and Chief Minister Pervaiz Elahi waving to people at the ceremony. 22 November 2004.

Public Transport in Lahore

Dharampura underpass, Lahore.

Hoardings encouraging investment in cars.

The Deteriorating Urban Environment in Lahore

the collaborative efforts of traffic engineering professionals from the World Bank, international and local consultants and TEPA. The project covered most of the primary and secondary road networks of the city. The main features of the project were channelization, junction improvement and enhancement of road capacity. The programme also contributed in introducing modern engineering practices for road construction in Pakistan.

Comprehensive Study on the Transportation System in Lahore (CSTS) (1990–2010)

A Comprehensive Study on the Transportation System in Lahore was carried out by TEPA with the technical and financial assistance of JICA in 1991. The study made various proposals for Lahore mainly developed through trip-generation, trip-distribution and traffic-assignment based computer models. The key projects suggested in this study included grade-separated road infrastructure at various junctions, the introduction of an urban rail system, utilizing existing intercity rail track and an LRT project.

Lahore Roads Rehabilitation Project (LRRP) 1997

This project was initiated in 1997 at a cost of 3.41 billion rupees financed by the provincial government as part of its annual development programme (ADP) in order to remodel the primary roads, and improve the surface of the secondary and tertiary road network in Lahore. Prior to this, the provincial government had made major contributions in funding road network projects carried out by TEPA and the communication and works department which were mainly making grade-separated underpasses along the Canal Bank Road, constructing overhead bridges on railway lines and building roads to the new terminal of the Lahore Airport.

Lahore Transport System (LTS) 1997

The Lahore Transport System (LTS) was created in 1997 to run public transport services by creating NGOs in collaboration with local private operators in order to follow the Faisalabad Urban Transport Society (FUTS) experiment. The LTS was registered with the provincial Social Welfare Department with funding arranged from private operators. This NGO was regulated by the law of social companies and administered by a governing body. The governing body typically comprised concerned government officers, community representatives, transporters and bus owners.

The model was developed on the basis of public-private-community participation to provide efficient public transport services in Lahore. The aim of creating this NGO was to generate funds for public transport through the private sector, renting existing infrastructure facilities and setting higher fares. The most interesting features of this NGO were the setting of their own fares (without approval of the government) and enforcement. Although this experiment was partially successful in the case of FUTS in providing efficient, reliable and decent public transport services by incorporating private and community sectors in the decision making process, lack of investment by the private and public sector in inducting new vehicles made this venture unsuccessful in Lahore.

Urban Public Transport Policy 1998

With the help of the World Bank and international consultants,[31] the Punjab government formulated a bus-based public transport policy which envisaged handing over the operation of bus services to franchisers. The goal was to phase out the chaotic 'wagon culture'[32] by introducing a regulated bus system owned and operated by the corporate private sector on the basis of route franchises.

The main role of the government was to regulate the services while the operational aspects were left to the private sector. Under this new policy, the government provided a package of incentives to attract private investment which included subsidized interest on bank loans, exemption of custom duty on import of CNG and diesel buses, and subsidized lease of the depot. In response to this policy, 708 new buses owned by nine corporate private groups started operating on thirty different routes in Lahore (information collected during field work in 2004). The policy was widely appreciated by public transport users due to the improved quality of the service. Although the bus-based public transport service in Lahore continues to run satisfactorily, the role of the government as a regulatory body remains questionable. Owing to a lack of institutional capacity of the Lahore Regional Transport Authority, the government has been unable to make long term plans for the development of the city's public transport system.

Integrated Master Plan for Lahore (2001-21)

The 'Integrated Master Plan for Lahore' was prepared for the future development of the city, from 2001 to 2021. All previous master plans prepared for Lahore favoured the urban road network, ignoring the potential of the railway network and the majority of trips that are non-motorized. The proposed 'Integrated Master Plan for 2021' is unabashed in its support for urban roads development. It seems that 'transportation' now only means road development.[33] Much of the discussion in the Plan document is under the heading of 'road network' which has clearly been given a high priority. This section proposes all short- and long-term investments be made on roads and road maintenance management, and on a central area traffic engineering and management scheme that would enhance the free movement of road traffic by increasing road capacity, parking and road safety. The subheadings on public transport, institutions, finances and the environment get very little attention in the Plan document. Finally, the first five-year programme for transport development in Lahore proposes to allocate 94.8 per cent of funding for road development, management and maintenance and only 5.2 per cent for a public transport terminal.[34] Clearly the plan is a 'road development plan', not, as it is called a comprehensive transport plan.

Both reports of the Integrated Master Plan 2001–2021 (Existing Scenario, Analysis and Proposal) totally neglected non-motorized transport, inspite of the fact that the CSTS in 1992 and the draft Integrated Master Plan of Lahore in 1997 found that non-motorized trips comprised 60 per cent of the total number of trips in Lahore. This neglect is evident from the fact that non-motorized trips found just one mention in the Plan document: 'Non-motorized transport, in particular vulnerable road users (i.e., pedestrians, bicyclists, etc.), are required to be mainstreamed by providing facilities for their mobility'.[35]

According to the Oxford Dictionary, 'mainstream' is defined as 'normal or conventional ideas, attitudes, or activities'. It would seem that by overlooking non-motorized transport in the Master Plan, the authors were subscribing to the traditional attitudes and ideas in the transport planning practices in Lahore. As a result and not surprisingly no money was allocated and no plan formulated to attain the one-sentence policy objective for non-motorized transport. It should be noted that the cost of building an entire network for non-motorized transport and its facilities is equivalent to building just a few kilometres of new roads. Even though it is acknowledged that the deteriorating environment of the city of

Lahore is due to the traffic situation[36] yet no measures were proposed to reduce this traffic. For institutional strengthening, a metropolitan transport authority has been proposed in the Plan, which is at least a step in the right direction, but whether appropriate and independent expertise can be found to staff it is another matter.

In conclusion, the Integrated Master Plan is an unconvincing piece of work for transportation. It is based on earlier American-style transportation studies, by now thoroughly discredited all over the world. This review of the proposals of the Integrated Master Plan 2001–2021 raises the question of why the Master Plan headed in the same direction as previous studies even though it was prepared with local financial and technical support. It would seem that the path set by earlier studies developed by international development institutions was firmly established and hence very hard to change. It is important to highlight the fact that the Integrated Master Plan 2001–2021 merely extended the road network proposed by LUDTS in 1980.

Urban Commission, Provincial Urban Transport Policy (PUTP) and City Development Strategy for Lahore[37]

The Punjab government prepared a 'Medium Term Development Framework' that was to be implemented during the period 2006–2009. The framework envisaged the establishment of an urban commission for preparing a comprehensive urban development policy. It was proposed to develop a provincial urban transport policy (PUTP) that would guide the future comprehensive urban transport strategy for Lahore. This policy was to be prepared with the technical and financial assistance of Cities Alliance, and was to be a part of the proposed Lahore city development strategy. The Alliance supports cities in preparing development strategies that 'link the process by which local stakeholders define their vision for their city and its economic growth, environmental and poverty reduction objectives, with clear priorities for actions and investments'.[38] The strategy was to be implemented by the proposed Punjab large cities development policy loan (DPL), with technical and financial assistance of the World Bank.[39] The objective of the proposed DPL was to complement the city development strategy for promoting economic growth in major cities of the Punjab through metropolitan level strategic planning, integrated infrastructure investment programmes and efficient urban service delivery. The improvement of urban transport was one of the key areas in the project.

Over 1 per cent of the province's total ADP budget for 2006–2007 was allocated to implement all these policies, plans and strategies. In contrast, the Punjab government allocated 40 per cent of the total ADP budget for road construction during the same period.[40] In addition, it committed an amount of Rs48500 million for road development to increase average road density from the 0.3 km to 0.4 km per sq km during 2006 to 2009.[41]

TRANPORT DECISION-MAKING IN LAHORE

Notwithstanding interventions from the powerful federal ministries and departments, it is the Punjab government and the LCDG who are primarily responsible for planning the city's transport and providing the associated infrastructure (shown in Figure 6.2).

The Punjab government's planning and development department (P&D) is the principal planning organization of the province. The designated role of the P&D is the preparation, coordination and supervision of the strategic provincial five-year and annual development programmes, and also the allocation of financial resources to various sub-sectors. The P&D has a roads and bridges section through which it ensures that the provincial transport policy and planning is incorporated into those made for Lahore. This section also co-ordinates incorporation of the provincial transport policy and planning components into the national five-year plan prepared by the federal Planning Commission. The P&D also has an environment division which incorporates the environmental agenda into the provincial transport policy. Despite its essentially strategic role, the P&D department has been involved in some specific provincial transport policy matters because it enjoys financial autonomy and acts as a coordinator with the federal Planning Commission.

The transport department in a province is its key organization for formulating transport policy; it is responsible regulating public transport and freight services in coordination with the provincial transport authority (PTA). The PTA in turn works with regional transport authorities (RTAs) responsible for defined administrative regions of the province to organize local route permits, transport fares, speed limits and parking for all passenger and freight vehicles. In Lahore, the Lahore Regional Transport Authority (LRTA) is responsible for public transport planning and implementation of related projects.

FIGURE 6.2

ORGANISATIONS RESPONSIBLE FOR TRANSPORT DECISIONS IN LAHORE

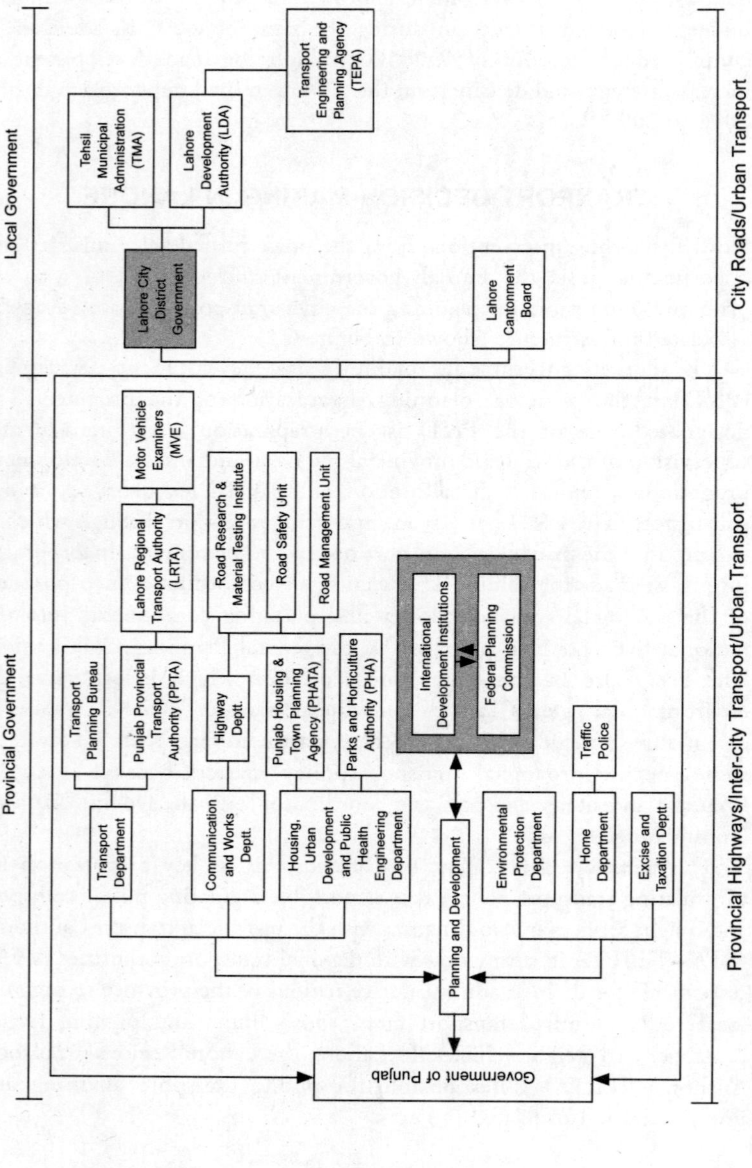

Source: Compiled by author

The communication and works (C&W) department, formerly the public works department (PWD), is responsible for the construction and maintenance of all provincial roads and bridges. A road research and material testing institute, and road management and road safety units have been established under C&W to provide technical and managerial support. The oldest road building organization in the province, the C&W department exercises great influence in road construction and associated infrastructure provision in the city of Lahore.

The Punjab environment protection department (EPD) is responsible for promoting sustainable development although in the case of Lahore, this is done in coordination with the district environment officer of the LCDG. The main task of the EPD and its subsidiaries—the EPA and the LCDG—are to conduct surveys to assess vehicular air and noise pollution and to impose penalties on polluting vehicles.

The provincial housing, urban development and public health engineering department is responsible for establishing a comprehensive system of planning and development in urban areas under the Punjab Development of Cities Act, 1976. Accordingly, development authorities like the LDA were created in all metropolitan cities of the Punjab to prepare and implement master plans for long term urban development. After the implementation of the Punjab Local Government Ordinance (2001), the LDA was transferred to LCDG. However, a Punjab housing and town planning agency (PHATA) was created in 2002 to formulate the provincial land-use plan, and the regional, master and outline development plans for urban and rural areas of the entire province.

The LCDG was created in pursuance of the Local Government Ordinance 2001. The LDA, a component organization of the LCDG, is responsible for formulating comprehensive metropolitan plans for Lahore, whereas the TEPA, also under the LDA, is responsible for preparing comprehensive transportation plans and urban transport policies for the city with complete authority to implement traffic engineering and traffic management programmes.

The primary responsibility for providing municipal works and services in Lahore is vested with the *tehsil* municipal administrations (TMAs) who work under the LCDG. The TMAs also manage street furniture and roadside parking, and may auction some on-street parking rights to the private sector without being required to consult the TEPA. The TMAs are usually funded from the revenue generated by property taxes in Lahore.

In summary, seven provincial departments and two local government departments are directly or indirectly responsible for transport related decision-making in Lahore. Their detailed responsibilities are shown in Table 6.1. Additionally, transport and road related funds from Lahore are collected by the federal government, the Punjab government and the LCDG. The federal government collects duties and fees accruing from the sale of fuel and lubricants, vehicles and vehicle parts. The NHA collects toll fees from traffic originating from and entering Lahore. The Punjab government's highway department (under the C&W department) has ownership of all other intercity roads which are not administered by the NHA and therefore collects toll fees from traffic originating from and entering Lahore using such roads. The Punjab excise and taxation department issues driving licenses and the fees collected under this head are deposited in the provincial exchequer. The traffic police collect fines from traffic violators and similarly deposit this money in the provincial exchequer. The Parks & Horticulture Authority (PHA) as an autonomous body exercises its right to collect fees from roadside hoardings and parking of vehicles in Lahore. The LCDG is responsible for collecting terminal fees from public transport operators, and roadside parking fees. Before the creation of the LCDG and PHA, the TEPA was responsible for collecting fees from roadside hoardings and parking of vehicles. In short, the departments involved in transport related decision-making in Lahore have great interest in either constructing roads or collecting charges from road users.

TABLE 6.1
ORGANIZATIONS AND THEIR RESPONSIBILITIES FOR URBAN TRANSPORT PLANNING AND DECISION-MAKING IN LAHORE

Punjab Government Departments	**Key Responsibilities**
* Planning & Development Department • Road and Bridges Section • Urban Development Section • Environment Section 1. Parks and Horticulture Authority (PHA)	* Provincial five-year development plan including transport * Finance allocation of roads projects within their own budget or request to federal government for financial help * Assessment of material and human resources of provincial transport institutions * PHA—collect fees for roadside hoardings and parking of vehicles in Lahore * – Environment protection
* Transport Department • Transport Planning Bureau (TPB) • Punjab Provincial Transport Authority (PPTA) 1. Regional Transport Authority (RTA) / Lahore RTA • Motor Vehicle Examiners (MVE)	* Formulation of transport policy * Urban transport planning * Traffic speed limits, loading regulations, parking permits * Payment of compensation in accidents * TPB—planning, monitoring and evaluation of road transport services in Punjab * – Conducting research to formulate transport policies and projects, executing and implementing transport policies and projects, developing and maintaining computerized transport data base * – Liaising with international institutions on the development of road transport sector * PTA—granting inter-provincial routes permits to freight and public transport operators * – Franchised bus operations permission in Lahore and other cities of the province * – Management of bus stands * (L)RTA - grant, issue and renew route permits to urban and intercity public transport within Lahore region * – Prepare time tables and fare tables for public transport, taking decision on concessional fare, grant license for bus stands, approval of bus stops * – Formulate regulation for motorcycle rickshaw * – Issue and renew licenses to vehicle body building workshops

	* – Checking traffic, traffic management, traffic survey and public transport data base * MVE - fitness examination or road worthiness of commercial vehicles * – Authorized to hold driving tests as a member of the driving test board * – Investigation of fatal accidents to give technical opinion
* Communication and Works Department (C&W) • Highway Department 1. Road Research & Material Testing Institute 2. Road Management Unit 3. Road Safety Unit	* Plan, design, construct, maintain, repair and improve provincial roads and bridges * Laying standards and specifications for various types of roads and bridges. * Testing of road construction material * Administration of roads and bridges, toll collection * Establishing of road network reference system and road assets management system * Carry out traffic count, condition & roughness survey
* Excise and Taxation Department	* Vehicle registration fee, transfer fee and annual token tax * Motor vehicle dealer registration
* Home Department • Traffic Police	* Enforcement of traffic laws and regulations * Operational aspect of traffic management such as road side raising and encroachments * Authorized to hold driving test * Road crashes information and data base
* Environment Protection Department • Environmental Protection Authority	* Survey and assessment of vehicular pollution * Evaluate initial environmental examination (IEE) and environmental impact assessment (EIA) of road and railway projects * Take measures to promote sustainable development and provide information on environment friendly technology. * Provide ambient air quality testing & monitoring facilities.
* Housing, Urban Development and Public Health Engineering Department • Punjab Housing and Town Planning Agency (PHATA)	* PHATA—Upgrading and developing big cities * – Preparation of a provincial land-use plan supported by regional and district land-use plans * – Regional development plans, master plans, outline development plans, small town development plans, district/tehsil/union council plans

Lahore City District Government	Key Responsibilities
* Lahore Development Authority (LDA) • Traffic Engineering and Transport Planning Agency (TEPA) * Tehsil Municipal Administration (TMA)	LDA * Responsible to prepare 'metropolitan development plan' and implement with 'annual development programme' * Prepare, implement and enforce schemes for environmental improvement (such as phasing out of *tongas*) and checking environmental pollution by reducing vehicular exhaust and congestion * responsible for beautification of the area according to prescribe environmental standards TEPA * Urban transport planning and policy for Lahore * Urban transport studies, research, traffic survey, road accident data collection and analysis * Traffic engineering, traffic impact analysis * Preparation, implementation and operation of traffic management schemes including vehicular speed, encroachment removal, size and location of advertising signs and hoardings, location and relocation of petrol stations and motor workshops, bus and truck stops, traffic signals, parking policy, parking fee * Road building and infrastructure improvement in Lahore including roads project planning and construction management and specification * Road maintenance, road patching and road surface restoration * Construct, maintain, upgrade and exclusively charge toll on roads, bridges, flyovers, underpass or such facilities within LMA * Pedestrians planning and infrastructure provision * Plan and design in collaboration with concerned agencies the public transport system for LMA including route network, priority measuring for highest possible efficiency of the mass transit system * Responsible for road safety, including road safety education programmes, training courses and awareness campaign TMA * Responsible for construction of streets in areas of their responsibility
Source: Compiled by through field work and other secondary sources	

There are seven Punjab government departments and two departments of the LCDG which are mainly involved in transport and infrastructure policy, planning and implementation in Lahore. The effectiveness of these local institutions in formulating transport policy and preparing plans will be assessed in the following sections by exploring institutional, financial and human resource capacities.

LAHORE URBAN TRANSPORT INSTITUTIONAL CAPACITIES

To assess if institutions in Lahore can independently formulate transport policy and plans or whether they depend on external assistance, it is necessary to evaluate the capacities of these organizations. These capacities can be categorized into human resource, financial and coordination competencies that are required for efficient transport decision-making. A study of these institutional capacities will also help in assessing which institutions are capable of delivering on their vision for transport in Lahore. By virtue of their financial powers and high professional skills, institutions usually control the pattern of transport planning in the city. These capacities were developed over time and represent the power of some institutions over others in the transport decision-making process. Finally, the discussion will permit us to make an inference whether international development institutions helped in developing capacities amongst local institutions in Lahore, and if so, which mode of transport and infrastructure development did this assistance favour.

I will explore the existing human resource situation in three provincial government organizations, i.e., planning and development (P&D), communication and works (C&W) and the transport department, and one local government organization, the TEPA, all of whom are directly responsible for transport planning and policy in Lahore.

The P&D is the principal planning department headed by a chairman, and assisted by members and a chief economist (see Figure 6.3). The designated role of the P&D is to coordinate and supervise the strategic provincial five-year plans. However, till the late 1980s which was when the transport department in the Punjab government was created, it was the P&D department through its member engineering (and now the project manager roads) who coordinated the provincial transport policy and planning components in the national five-year plans. The project manager roads is placed directly below the chairman P&D and is assisted by the chief of roads and bridges (R&B). There are four assistant chiefs

FIGURE 6.3

ORGANOGRAM OF PLANNING & DEVELOPMENT DEPARTMENT, GOVERNMENT OF PUNJAB

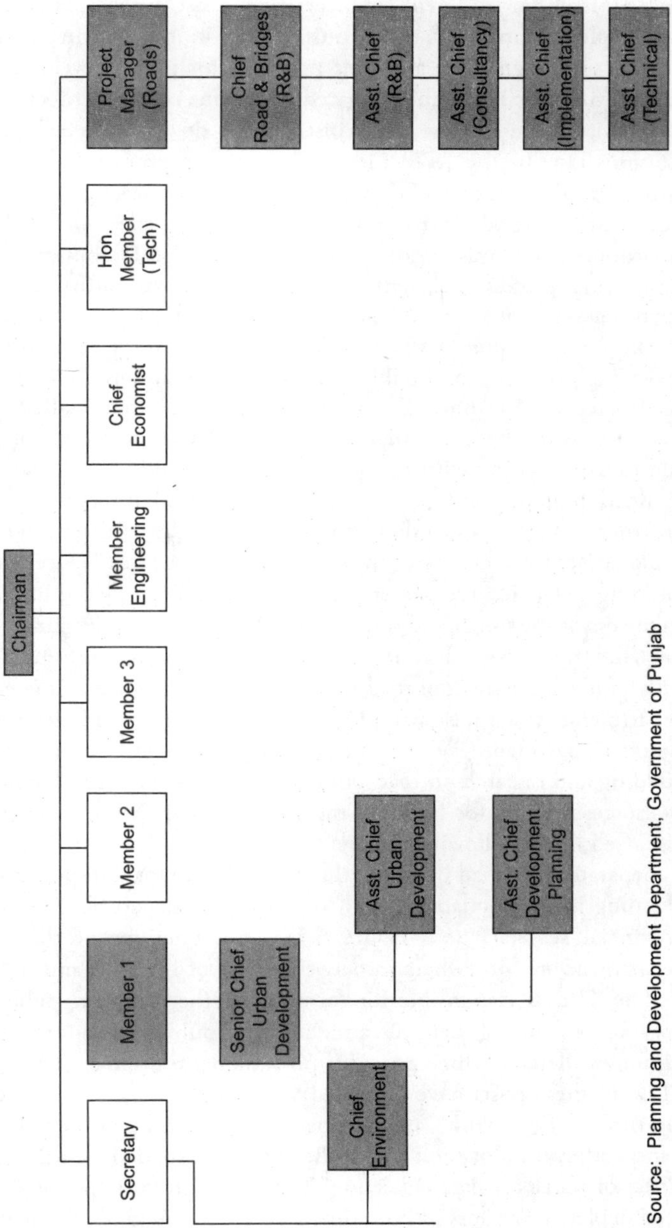

Source: Planning and Development Department, Government of Punjab

(R&B), mainly road (civil) engineers, who assist in road planning and in the implementation of road construction in the province, including Lahore. Although there are three positions for urban development, one chief and two assistant chiefs, placed below one of the members of P&D, economists occupy these posts just as they do in the federal Planning Commission. In the ADP for 2006–7, it was proposed to establish a 'urban unit' in the P&D to provide institutional support in urban development and management.[42] Recently, the post of a chief environmentalist was created under secretary P&D, but without any supporting professional staff it is doubtful if this office can achieve anything substantive. In the absence of appropriate human resources in the urban development and environment sections, the responsibility for reviewing pedestrian and public transport proposals has been given to the R&B section. In short, the P&D is almost entirely staffed by road engineers who play a dominant role in local transport decision-making and in coordinating with national level institutions.

Road building and provisioning of associated infrastructure in the province is the responsibility of the Punjab highway department which works under the secretary of the C&W department (see Figure 6.4). Road building activities are sectorized and accordingly placed under chief engineers of the north and south zones. Lahore falls under jurisdiction of the chief engineer north zone—specifically a superintending engineer (SE) with a full complement of road engineers. As one of the oldest government departments which was established before partition [then called the Public Works Department (PWD)], it has the human resource capacity for road building and research on road-building materials. The C&W department is also responsible for building the Ring Road in Lahore which will be discussed in the following chapter.

Separate from road building, the responsibility for transport policy and planning in the Punjab lies with the transport department headed by a provincial secretary (see Figure 6.5). The functions of the transport department are somewhere in between those of the TPB and the PTA.

The TPB is responsible for formulating the transport policy of the province including Lahore, and also for public transport planning. Although there are three posts for professional transport planners in the TPB, all these posts have remained vacant for a long time. In 2004 (at the time of field work), no member of the department was a qualified transport professional and few of the staff had experience in the technical aspect of transport data collection. The TPB's professional capability has been depleted by a loss of both functions and resources and it is therefore

FIGURE 6.4
ORGANOGRAM OF COMMUNICATION & WORKS DEPARTMENT, GOVERNMENT OF PUNJAB

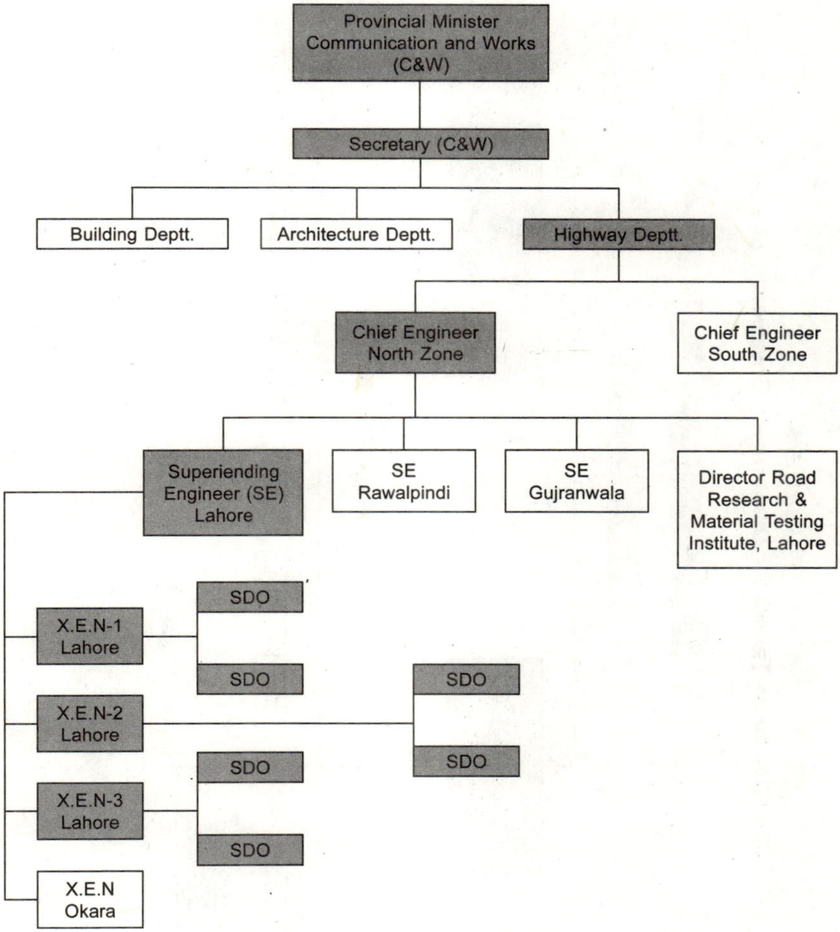

Source: Communication & Works Department, Government of Punjab

FIGURE 6.5

Organogram of Transport Department, Government of Punjab

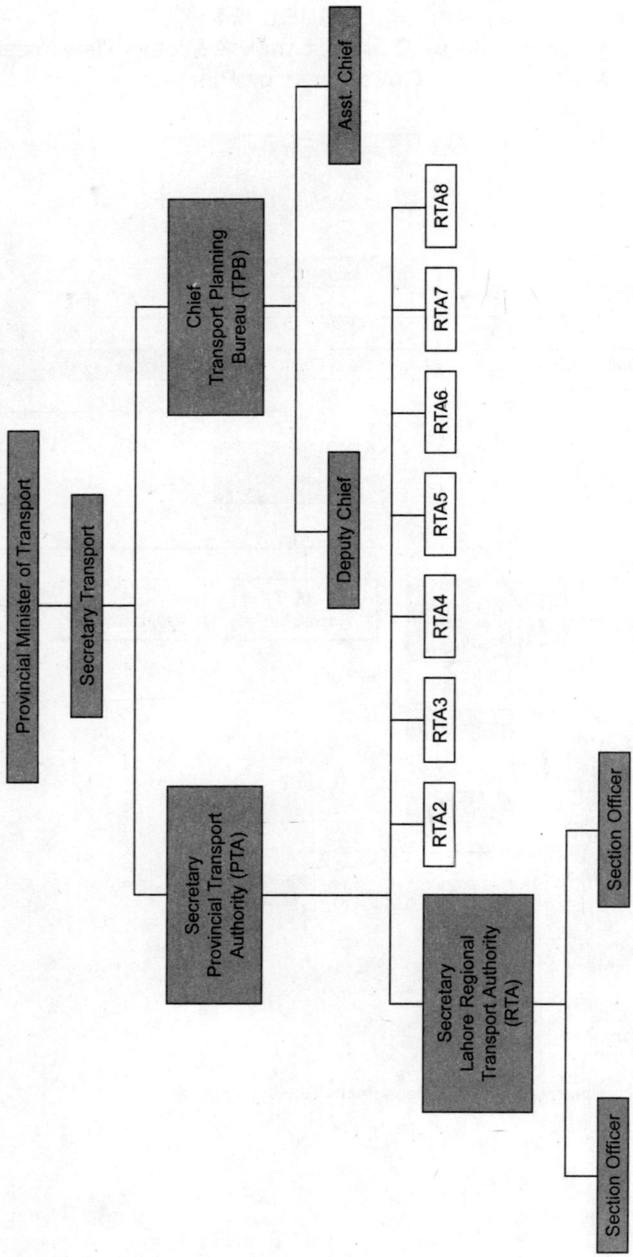

Source: Transport Department, Government of Punjab

unable to develop a comprehensive transport policy.[43] On the other hand, Lahore RTA regulates public transport by issuing route permits for passenger services and freight transport. In the absence of a transport professional, the decision to issue permits is taken on an *ad hoc* basis by the bureaucratic head (secretary) of the LRTA. Clearly, because of the current paucity of human resources, the transport department is unable to guide, innovate and expedite transport planning in the province, especially in Lahore. Despite the fact that there are a very limited number of technical and managerial staff positions, the transport department needs to give immediate attention to fill the posts of professional transport planners. Moreover, it needs to arrange up to date training for existing field staff on how to obtain advance data for public transport planning.

To compound this serious deficiency of human resources, the control that the federal government's establishment division has over the placement and promotion of personnel at both the provincial and local levels does little to alleviate the situation. The human resource capacities of provincial and local transport institutions consequently continue to decrease.[44]

In contrast to the serious deficiency of professionals in transport institutions in general and transport planning institutions in particular, the position in the Punjab highway department (of C&W) is relatively better. Although the road engineers of the C&W are well recognized for their professional competence because there are no transport planners in the P&D and the capacity of urban development personnel is weak, there is little understanding of the need for sustainable transport planning in Lahore. Also, the weak capacity of the transport department has enabled the P&D department to get involved in transport policy matters and act beyond its role of a strategic planning agency.[45] More specifically, the P&D played a leading role in discussions with potential bidders for provision of bus services in Lahore and in negotiating terms on which bus operations were to be franchised. This is more properly a function of the transport department as described earlier, but because the P&D department has a position of dominance due of its stronger and consistent capacity it arrogated to itself the role of a decision-making organization.

These facts are supported by a review of the finances allocated for road construction, public transport and the environment. The Punjab government allocated Rs8.3 billion for the construction and maintenance of the road network in 2004–5.[46] During the course of the year, the government allocated an additional Rs6.3 billion thereby investing a total of Rs14.6 billion for roads in that period. The LMA received Rs2.6 billion for the development of roads.[47] A large proportion of the money for roads was spent through the C&W department. Then again in the period 2005–6, the government allocated Rs12.5 billion for the construction and widening of roads, and for making flyovers and underpasses and an additional allocation of Rs5 billion was expected to be made during the year. In contrast, there is no money allocated for public transport, pedestrian and non-motorized transport planning. It was assumed that the private sector would invest in public transport but no policy was formulated to attract private investment into the public sector. In terms of the environment, Rs1 billion was allocated to establish a 'green fund' to reduce urban air pollution by investing in mitigation technologies in the transport sector.[48] However, detailed modalities and a timeframe to achieve these objectives were not defined.

Similarly, the financial strength of the C&W and the transport departments affirm the priorities of the Punjab government for road building. For example, it allocated a mere Rs2.1 million as development budget for the transport department in 2005–6[49] which was to be mainly used for the maintenance of offices. Additionally, Rs7.29 million was allocated as a non-development budget to be mainly used for paying staff salaries. In contrast, just the development budget of the highway department (C&W) was Rs3.99 billion during 2005–6. There is clear evidence that in terms of human resources and financial capacities, policies are patently biased in favour of road building organizations in the Punjab.

Institutional capacities at the provincial level have strongly influenced the functioning of local organizations. Under the LCDG, a subsidiary of the LDA, the responsibility for road engineering, traffic management, transport policy and public transport planning within Lahore is assigned to the TEPA. Despite this, TEPA's main focus has been on road building activities. From 1988 to 1997, it was the implementing agency for the World Bank-financed PUDP which was primarily concerned with road and junction improvement works. During the project, the TEPA human resource capability largely comprised of road engineers. The TEPA's technical team is headed by a chief engineer who is assisted by a director of engineering construction and a director of traffic engineering (see Figure 6.6). The agency has a reasonable number of road engineers which include six deputy directors and five assistant directors. Since the TEPA has mostly been concerned with road projects, its human resource capacity for urban, public and non-motorized transport planning is negligible. The agency can be credited for just one 'Comprehensive Transport Study for

FIGURE 6.6
ORGANOGRAM OF TRAFFIC ENGINEERING AND TRANSPORT PLANNING AGENCY, LDA

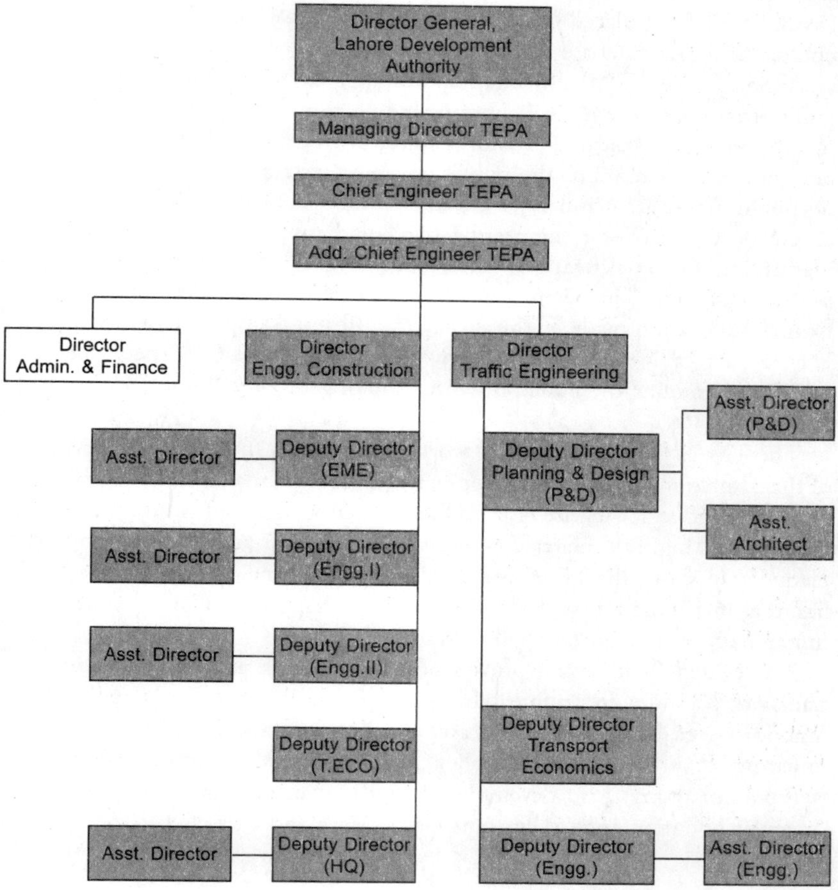

Source: TEPA, LDA

Lahore' which too was completed with technical assistance of the JICA. The twenty-year transport master plan was produced mainly by the JICA technical experts in 1992. Because the TEPA lacks requisite professional capabilities, it has not been reviewed for the purpose of this study; reliance has therefore been placed on JICA's studies to estimate trip demand and obtain the traffic database for Lahore.

Although the TEPA also has responsibility to 'plan and design the public transport system for Lahore including route network', it focuses mainly on road construction and at times on traffic management, and does not plan or monitor the public transport system. At present it has no public transport planners or experts. Meakin presented the TEPA as an entity with a strong engineering bias[50] and supported this claim by identifying the qualifications and experience of the agency's senior professional staff. This claim can be further verified by examining the financial allocation made by the agency to different transport activities in Lahore (see Table 6.2) which clearly shows that the TEPA spent the maximum amount of money on road building activities over the past many years.

The TMAs provide municipal works and services in Lahore. As part of their functions related to development control and management of the municipal services, they are responsible for managing street furniture and roadside parking. But there are no posts in their organization for transport planners. In short, the TEPA and the TMAs have some capacity for road building in Lahore but without professional transport planners, they are functionally and organizationally seriously deficient.

It is evident that several provincial and local organizations work on transport planning in Lahore with many of them having overlapping functions (see Table 6.3). The areas of fragmentation relate to the functions of urban transport policy making, spatial planning, public transport planning, road development, traffic management, road safety, vehicular pollution and urban transport research and data collection.

TABLE 6.2
TEPA FINANCIAL EXPENDITURE ON DIFFERENT TRANSPORT ACTIVITIES

Years	*Roads, Flyovers, Underpasses Roads, Land Acquisition and Road Feasibility Studies*	*Public Transport/ LRT*	*Traffic Data/ Traffic Surveys*	*Traffic Management/ Area Traffic Control, Signals*
1993–94	66.15	1.5	0	30
1994–95	144.83	0	0	10.93
1995–96	170.06	0	1.5	56
1996–97	82.68	0	0	11.4
1997–98	133.25	0	0	11.5
1998–99	192.53	5	30	0
2000–1	331.73	1	0	0
2001–2	50.72	0	0	0
2002–3	65.4	0	0	0
2003–4	47.67	0	0	0
2004–5	305.98	0	0	0

Source: Figures taken from TEPA budget estimates of several years
* Data is not available for 1999–2000

The urban transport policy in Lahore is the responsibility of the TPB, the transport department and the TEPA. The P&D has the responsibility for incorporating national transport policy goals at the city level. However, because of lack of human and financial resources for transport planning, no policy for Lahore has been prepared thus far. However, at the national level the Planning Commission and the NTRC have formulated transport policies including some general guidelines for urban areas in Pakistan. Transport policies are effective only in conjunction with spatial planning at the local level. Although the LDA and PHATA are responsible for spatial planning in Lahore, the latter has limited its role to housing schemes developed before the creation of LDA in 1975, and has now shifted its focus on to other cities of the Punjab.

TABLE 6.3
INSTITUTIONAL FRAGMENTATION IN TRANSPORT PLANNING IN LAHORE

Key Principles	Department Responsible
* Urban Transport Policy • Urban Transport Planning	* Planning Commission, Govt. of Pakistan * NTRC, Govt. of Pakistan * P & D, Govt. of Punjab * TPB, Transport Department, Govt. of Punjab * TEPA, LDA, Lahore City District Govt.
* Spatial Planning • Metropolitan Plan • Land Use Plan • Department Plan	* LDA, Lahore City District Govt. * PHATA, Govt. of Punjab
* Public Transport Planning • Route allocation/ Timetable • Bus terminal • Fares decision/ concession	* TPB, Transport Department, Govt. of Punjab * Lahore RTA, PTA, Transport Department, Govt. of Punjab * TEPA, LDA, Lahore City District Govt.
* Private Mode of Transport Planning • Car growth planning • Motorcycle growth planning • Parking policy	* Transport Department, Govt. of Punjab (Parking permits) * Excise and Taxation Dept., Govt. of Punjab (Car and M.C registration fee) * PHA, Govt. of Punjab (collection of parking fee) * TEPA, LDA, Lahore City District Govt.(Parking policy)
* Non-Motorized Transport Planning and Infrastructure Provision • Pedestrian Planning • Cyclist Planning • Horse Driven Vehicles	* TEPA, LDA, Lahore City District Govt.
* Traffic Management Schemes • Speed • Traffic Signals	* TEPA, LDA, Lahore City District Govt.

* Road Infrastructure Planning and Development • Road construction • Underpasses, flyovers, interchanges and constructions and designs * Administration of road, bridges toll	* NHA, Govt. of Pakistan * Highway Department, C&W, Govt. of Punjab * TEPA, LDA, Lahore City District Govt.
* Road Safety	* Transport Department, Govt. of Punjab (payment of compensation in accidents) * Traffic Police, Home Department, Govt. of Punjab * TEPA, LDA, Lahore City District Govt.
* Vehicular Pollution • Air Pollution Monitoring • Noise Pollution	* EPD, Govt. of Punjab * PHA, Govt. of Punjab * LDA, Lahore City District Govt.
* Urban Transport Research • Research Studies/ Vehicular Pollution Studies • Data Collection/ Traffic Survey/ Accident data • Transport education/ campaigns	* NTRC, Govt. of Pakistan * Pak-EPA, Govt. of Pakistan * TPB, Transport Department, Govt. of Punjab * LRTA, Transport Department, Govt. of Punjab * Highway Department, C&W, Govt. of Punjab * Traffic Police, Home Department, Govt. of Punjab * EPA, Govt. of Punjab * TEPA, LDA, Lahore City District Govt.
Source: Compiled by author through field work and secondary sources	

Public transport planning in Lahore is a responsibility that is shared by the TPB and the Lahore RTA in the transport department and the TEPA under the LDA which is part of the LCDG. Similarly road infrastructure planning and development is the responsibility of both the Punjab highway department under C&W and the TEPA. However, in the absence of a clear policy for the development of public transport, combined with a serious deficiency of human and financial resources, neither the transport department nor the TEPA is currently engaged in public transport planning. These capacity deficiencies create uncertainties regarding the government's commitments to promote public transport in Lahore and as a result the private sector is discouraged from investing. In addition, laws and rules for public transport planning are inadequate and outdated with only The Motor Vehicle Ordinance 1965 and The Motor Vehicles Rules 1969 currently providing the legal authority to the LRTA to oversee the building (of bus bodies), operation and licensing of buses. These laws require revision to make them relevant to changed circumstances.

The TEPA and C&W compete with each other for road development in Lahore. For example, several road underpasses and some sections of the Ring Road in Lahore were built by the TEPA but are now being taken over by C&W. Additionally, the NHA also competes in road building in Lahore when national or foreign money is involved. In terms of road safety, vehicular pollution, urban transport research and data collection for Lahore, the following agencies are involved: the NTRC and the Pak-EPA at the national government level, the transport department, traffic police, EPD, C&W and the PHA at the provincial government level and the TEPA and LDA at the city district government level. However, these organizations limit their work to day-to-day activities rather than making a coordinated effort to address road safety and vehicular pollution issues in an integrated manner.

The fragmentation of urban transport planning responsibilities amongst several government departments and the absence of effective coordination mechanisms are at the root of Lahore's urban transport problems. A district transport advisory committee was created in 1989 to effect coordination amongst all entities responsible for transport policy and operations in Lahore. However, by including representatives of private transport operators and associated associations the committee has become more a mediator between private operators and the government and less a body that coordinates government transport policy. Institutional fragmentation among transport institutions in Lahore was also found by Meakin in 'Lahore Public Transport Policy and Institutions Study' as a part of the Lahore Urban Transport Project in 1998.[51] Meakin identified fragmentation of urban transport responsibilities, absence of effective coordination mechanisms, lack of monitoring and evaluation and an approach based on solving urgent day-to-day problems in several transport institutions in Lahore.[52] In the past, new organizations were created to fulfil roles and tasks that might well have been assigned to existing institutions by objectively evaluating their performance and appropriately developing their capacities. This was not done and as a result responsibility for transport planning in Lahore has been split amongst several organizations. The situation is further exacerbated due to neglect of the transport planning sections in these organizations; they require immediate attention to improve their human resource and financial capacities so that as a very minimum they are brought at par with the road building organizations in Lahore. Moreover, to achieve the goal of sustainable urban transport planning in Lahore it is necessary to have a central mechanism to coordinate activities of all relevant organizations.

Based on the discussion thus far one can argue that in the absence of qualified personnel, lack of financial resources and even basic office equipment and amenities, it is impossible to develop a framework of policies and plans for the transport sector within which organizations can coordinate and set sustainable transport planning priorities in Lahore. Transport activities at the local level only work with the help of provincial departments and agencies. This working relationship is occasionally observed amongst the TEPA, C&W and P&D, mainly for road building in Lahore; the relationship exists because these institutions have higher financial and human resource capacities for road building. However, similar relationships are not established for transport policy making and transport planning.

In the absence of appropriate capacities at the provincial and local levels in Lahore, the federal government exercises control of finances (budgeting, revenue, financial management) and personnel management of (staffing, reassignments, and promotions) over the provincial and district governments, as is the case in Karachi.[53] This pattern of administrative control is a continuance of the practices established by the British in pre-independence India.[54] The federal government relies on the advice of the international development institutions for urban transport projects in Lahore. Although there can be argument about the excellence and professionalism of these institutions and associated foreign consultants working in Pakistan, it must be pointed out that they are handicapped by lack of local knowledge and issues, and are short on planning experience in the Pakistani environment. In certain projects, international consultants involved local experts or government organizations but their roles were limited to collection of data and informal discussions. No serious effort was made by the federal or international institutions to build appropriate capacities of existing transport institutions so that could independently design indigenous solutions.

Establishing a sustainable transport system demands flexible, integrated and inter-modal transport development skills that can incorporate a range of factors like environment, energy, social welfare and urban development into the process of transport decision-making.[55] Therefore, institutional capacity building for sustainable transport must go beyond retaining economic or road building professionals. It must recognize the importance of multi-disciplinary knowledge, skills and people and their role in restructuring choices for delivering an urban transport system. It must also respond to the decision-making structure within which transport investment, management and coordination amongst all the various organizations and stakeholders take place. Obviously, this will be a complex and multi-layered capacity building programme in local and provincial organizations.

NOTES

1. Government of Pakistan, Federal Bureau of Statistics 1998.
2. Khan 1996: 507.
3. Govt. of Pakistan, EUAD and IUCN 1992: 75; ADB 1993.
4. http://www.statpak.gov.pk/depts/pco/statistics/area_pop/area_pop.html
5. Extracted from LDA 2004a: 1–2 to 1–7.
6. Growth rate extracted from Table 3.3, LDA 2004a: 3–3.
7. Information obtained from the Lahore City District Government website.
8. LDA 2004a: 6–7.
9. Ibid.
10. Ibid.
11. LDA 1997: 6–3.
12. Information provided by the Transport Department, Govt. of Punjab.
13. LDA and World Bank 1981c.
14. LDA 1997.
15. Estimated from Table 20.24 in Pakistan Statistical Year Book, 2007 available at http://www.statpak.gov.pk
16. LDA 1997.
17. LDA 2004a; Shigeta 2002.
18. World Bank 2006c.
19. LDA 1997: 13–22.
20. Ibid., 6–25.
21. Goulding, 1924: 47.
22. Glover 1999.
23. Dina 1899; Goulding 1924; Glover 1999.
24. Glover 1999.
25. *Daily Jang*, 25 July 2004.
26. Glover 1999.
27. Ibid.
28. Marchall 1972: 383.
29. Glover 1999.
30. Volvo 1980.
31. Meakin 1998.
32. A 'wagon' is usually a Toyota Hiace, an 18-seater, low-height vehicle.
33. See Integrated Master Plan 2001–21: Chapter 21.

34. LDA 2004b: 21–7.
35. Ibid., 21–1.
36. Ibid., 21–19.
37. These are proposed projects contained in the Punjab provincial budget 2006–7 announced on 14 June 2006.
38. Cities Alliance 2006:1, http://www.citiesalliance.org/doc/resources/cds/cds-guidelines/cds-gudelines-final-march06.pdf
39. World Bank 2006b.
40. Govt. of Punjab 2006: 42.
41. Ibid., 55.
42. Govt. of Punjab 2006: 50.
43. Meakin 1998.
44. LaPorte 1992: 108–112.
45. Meakin 1998: 7.
46. Govt. of Punjab 2005a: 39–40.
47. Govt. of Punjab 2005b: 95.
48. Govt. of Punjab 2005a: 42.
49. www.punjab.gov.pk
50. Meakin 1998.
51. Ibid.
52. Ibid.
53. Hasan 2005.
54. LaPorte 1984: 246.
55. Imran and Low 2003.

7

The Lahore Ring Road: Exploring Path Dependence

The focus of this chapter is on the historical path of the Lahore Ring Road, one of the emblematic transport projects in Lahore. The historical evolution of the project will help to explore whether the national critical juncture (1955–65) had an impact on urban development. The chapter examines the Ring Road project in Lahore as a potential outcome of a path-dependent transport policy at the national and local levels. This case study will analyse how planning, financing and construction decisions were made for this project. The study will further evaluate how the technical and financial principles of international development institutions influenced the policy-making discourse and practices in Lahore.

This chapter first traces the evolution of the Lahore Ring Road alongside a description of the project. The surrounding discussions and an analysis of the relevant international, federal, provincial and local institutions will be conducted in the above context. The subsequent section will analyse the nature of the technical arguments forwarded in justification of the Ring Road. In explaining storylines in the next section, the focus will be directed at the discourse field to assess whether there is discursive path dependence in urban transport policy. Finally, the Ring Road discussion is compared to the findings of Chapters 3, 4 and 5, and conclusions are drawn as to the nature of path dependency in transport planning, both locally and nationally.

EVOLUTION OF THE RING ROAD PROJECT IN LAHORE

The Lahore Ring Road has retained a position of importance as a transport infrastructure project for the last forty years. Describing the project as it evolved through various stages will be the main aim of this section. The study has categorized the evolution of Lahore Ring Road into four distinct decades; 1960s, 1980s, 1990s and 2000s.[1]

FIGURE 7.1
Ring Road in Different Master Plans

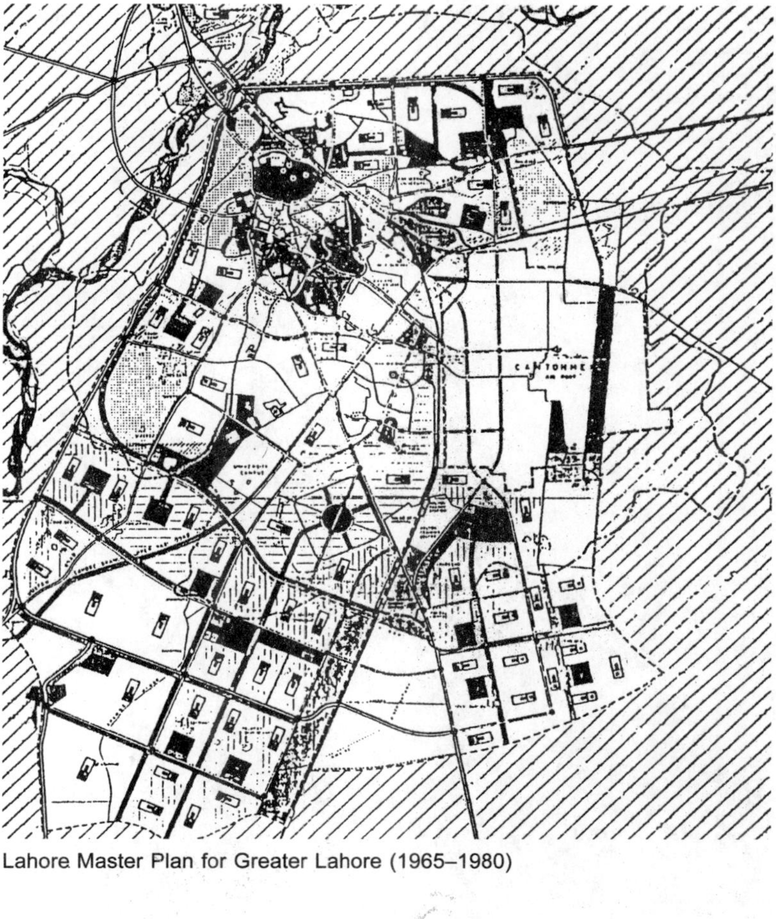

Lahore Master Plan for Greater Lahore (1965–1980)

FIGURE 7.1 (continued)

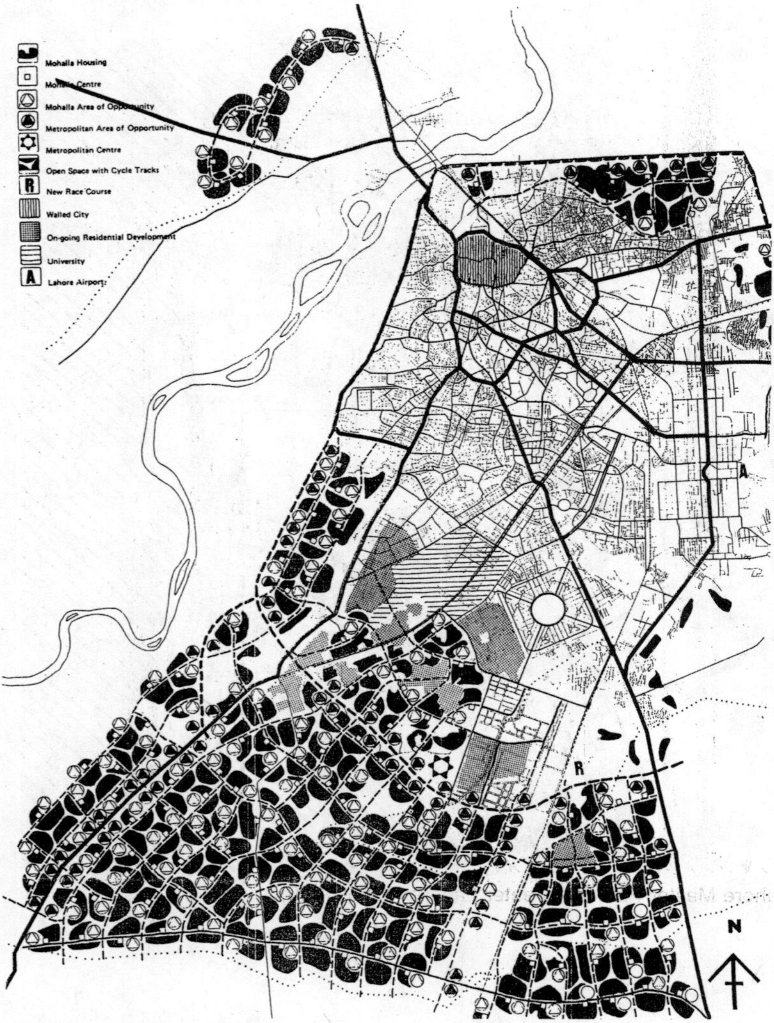

Lahore Structure Plan (1980–2000)

FIGURE 7.1 (continued)

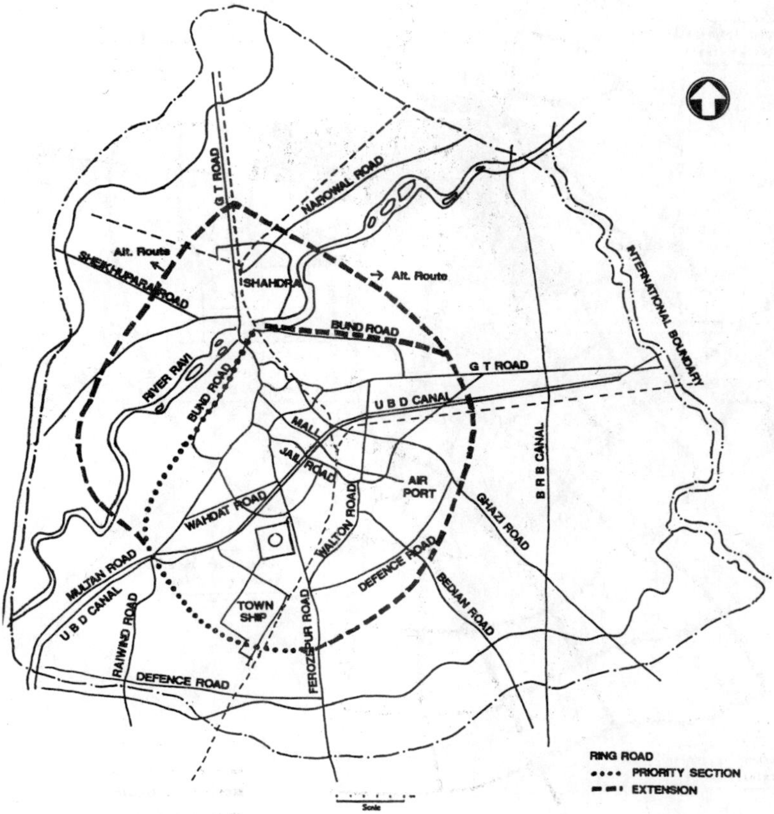

Punjab Urban Development Project (1988–97)

FIGURE 7.1 (continued)

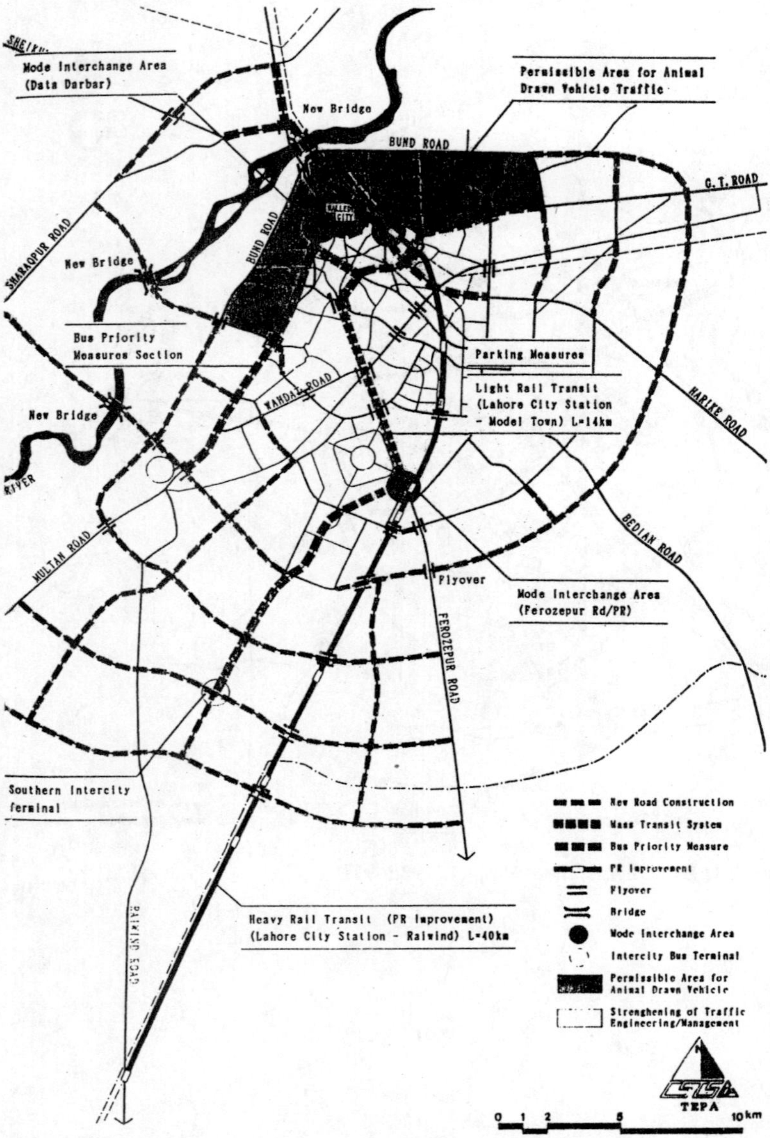

Comprehensive Study for Transportation System in Lahore (1990–2010)

FIGURE 7.1 (continued)

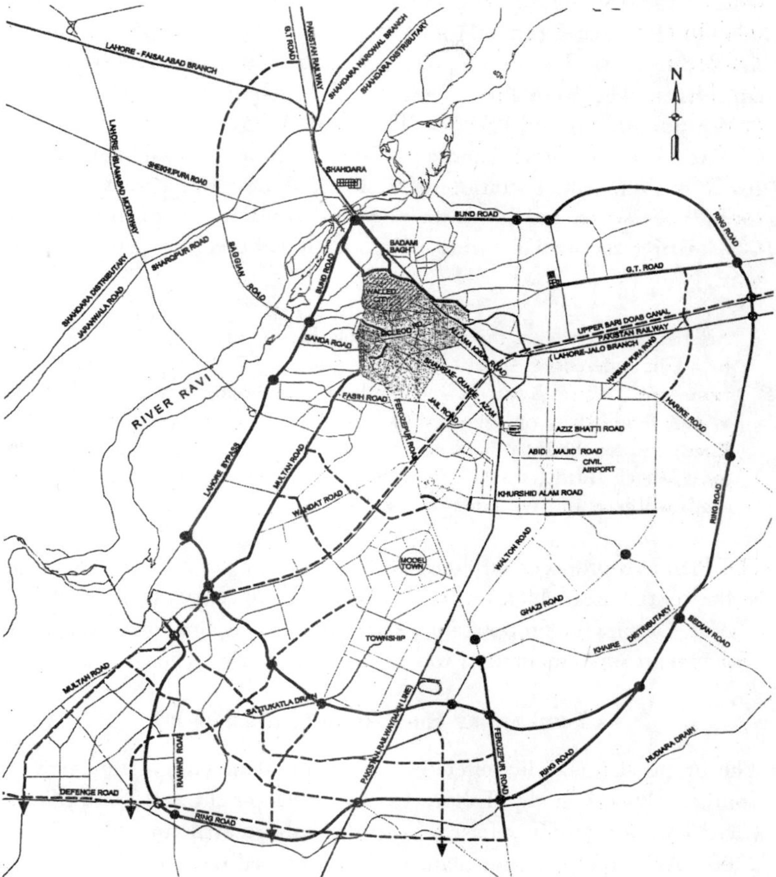

Integrated Master Plan for Lahore (2001-21)

The Lahore Ring Road in the 1960s

The Lahore Ring Road was initially identified as a 'circumferential arterial road' in the first Master Plan for Greater Lahore to be implemented from 1965 to 1980 (see Figure 7.1). The Plan was the first attempt to rationalize development in Lahore. It was prepared with the help of foreign consultants who were already engaged for preparing the country's First (1955–60) and Second Five-Year Plans (1960–65).

The Plan noted that Lahore had developed along four main arterial roads but the concentration of residential development was mainly towards the south and south-west because of a flood plain in the north. Considering the radial nature of roads in Lahore, the Plan proposed that:

> Arterial roads entering the city in a radial fashion are proposed to be joined by a Circumferential Arterial Road. The proposed circumferential road starts from mile 11.2 on Kasur Road (now Ferozupur Road), crosses Multan Road at mile 9 and there on becomes a part of New Multan Highway (now Bund Road), crosses the Ravi Road near the existing road bridge on River Ravi and joins Wagah Road (now GT Road) on mile 7. All the junctions with major roads will have to have grade-separation.[2]

The Plan also proposed an 'inner Ring Road',[3] perceived as a bypass road in the central area which was to circumscribe the central business district (CBD). Despite the proposal for a 'circumferential arterial road', no serious effort for its implementation was made in the period 1965-1980.

The Lahore Ring Road in the 1980s

The proposal for a circumferential road was retained and renamed the Southern Bypass in the 'Lahore Urban Development and Traffic Study' (LUDTS), commonly referred to as the Lahore Structure Plan (1980–2000). As in the past, this plan too was prepared by foreign consultants with limited input from local experts. The plan was financially supported by the World Bank.

The Southern Bypass or Ring Road appeared in the 'Urban Planning Report' which proposed that:

> The existing bund road should be extended in the future south-westwards, parallel to Multan Road....and then turn south-eastwards, crossing Multan Road and the Canal Road to pass south of Lahore Township and the new racecourse before reaching Ferozepur Road.[4]

The proposed alignment was made in accordance with the recommendations of the structure plan, with emphasis on the south and south-west growth of Lahore. However, the 'Traffic Study Report' of the plan limited itself to the central area traffic management.

Some other efforts to propose, justify and give impetus to the Ring Road project were made in the 1980s. In the early part of this decade, the National Highway Board (NHB), later named the National Highway Authority (NHA), conducted a feasibility study for building a bypass or a peripheral road around Lahore to develop long distance traffic. However, the Punjab government rejected this proposal because it did not serve the needs of Lahore.[5] The inception report of the Lahore traffic and transport component of the Punjab urban development project in 1988 included the Lahore Ring Road proposal. The report did not agree with the argument that the Ring Road would be beneficial for long distance traffic but acknowledged that it could be useful in distributing trips to the periphery of the city.[6] In 1988–89, the National Engineering Services Pakistan (NESPAK) produced a project concept (PC1)[7] to align the Ring Road with a new bridge on the River Ravi. However, it was realized that local problems could not be solved in the absence of a transport master plan. This realization resulted in the preparation of a Comprehensive Study on Transportation System (CSTS) in Lahore in 1990 undertaken by the JICA.[8]

THE LAHORE RING ROAD IN THE 1990s

The Lahore Ring Road received attention in the CSTS which was to be implemented during the period from 1990 to 2010. This was the first transport master plan prepared for Lahore with the technical and financial support of JICA. It supplemented the proposal for the Ring Road and recommended an extensive highway network with allied infrastructure. It proposed that a new 45.5 km circumferential road named the 'Outer Ring Road' be built to connect the existing Bund and Sharqpur Roads; this also required building a new bridge on the River Ravi. [9] The proposed length of the Outer Ring Road was additional to the 8.25 km of road which connects the existing Bund Road to the Ring Road.[10] In short, 53.75 km of new road along with up-gradation of the existing Bund Road was proposed as a 'Ring Road' in the CSTS.

On the recommendation of the CSTS and with finance made available from the World Bank, a Lahore Ring Road feasibility study was carried out by international consultants in 1992. The study proposed an approximately 65 km-long Ring Road encircling an area that lay within a diameter of around 20 km from the city of Lahore.[11] It was argued that the main role of the Ring Road was to provide circumferential access to disperse traffic and relieve radial roads.[12] The study categorized the Ring Road into various sections according to their functions. It was suggested that the section from Babu Sabu to Multan Road (southern section) would in addition serve as a strategic route to facilitate intercity traffic between the north and south corridors of Pakistan. The study envisaged the southern section as a 'primary distributor', with characteristics such as no pedestrian or frontage access, no parking and with vehicular speeds of not less than 80 km/hr; hence mainly designed for free and fast moving traffic.[13] It was also perceived that little access to the Ring Road was to be expected from the east because of the political situation between India and Pakistan. In light of these feasibility reports, the TEPA started to acquire land for the Ring Road as part of Phase 1 in 1995.

Beside the efforts of the provincial and local governments to finalize details of the Lahore Ring Road, federal government departments also started to take interest in building some parts of the Ring Road according to their requirements. At first, the NHA conducted a feasibility study on ways to improve the National Highway (N-5) section near Lahore in 1991. The objective was to improve access and traffic around the Lahore Metropolitan Area. The study suggested that improving the existing Bund Road (part of the Ring Road) would improve traffic conditions and was economically justified.[14] Similarly, the federal government started construction of the Lahore-Islamabad Motorway in 1991. Lahore was connected to the 'zero point'[15] of the Lahore–Islamabad Motorway by building one section of the Ring Road known as the Lahore Bypass via a new bridge on the River Ravi. The project was opened for traffic in 1997. The Lahore Bypass was also conceived as a means to give through traffic an alternative route to the Lahore central area.[16]

Although one section of the Lahore Ring Road was built by the federal government, the Punjab government along with the TEPA were interested in building the complete Ring Road in accordance with the recommendations made in earlier studies. The Punjab government's efforts met with success in 1999 when the World Bank showed an interest in funding the proposed Lahore Urban Transport Project (LUTP) which aimed at expanding and improving the road infrastructure in the Lahore Metropolitan Area. High priority road links such as the Lahore Ring Road and grade-separated junctions were important components of the project. Despite its initial interest, the World Bank subsequently dropped the project because of a change in the country's political situation. Later in 1999 the Punjab government again initiated attempts to build the Ring Road, this time with private sector participation on a BOT (built, operate and transfer) basis. Messrs Daewoo showed interest in the project and engaged the same consultants to work on different models and various alignments. However, because of a change in the alignment of the Ring Road the project met with resistance on account of some political and economic considerations. The proposal did not mature during this decade.

The Lahore Ring Road in the 2000s

After the Lahore Structure Plan (1980–2000), the Integrated Master Plan for Lahore was prepared by local consultants for the period 2001–2021. This plan formulated a 'Road Development Policy and Strategy' for the Lahore Metropolitan Area. The policy strongly recommended the early construction of the entire Ring Road as a result of which this project was placed in the short-term investment programme. It proposed to allocate 90 per cent (Rs16, 000 million) of the total transportation sector budget of the five-year programme towards completion of the Lahore Ring Road.[17]

At last, after forty years of various proposals, formal construction of the Ring Road started on 22 November 2004 at a ground-breaking ceremony attended by the President of Pakistan. The construction started with the building of a road interchange at Niazi Chowk. It was expected that the interchange would be completed by the middle of 2006. The C&W department of the Punjab government is currently responsible for the Ring Road project, while NESPAK is preparing detailed feasibility studies in association with M/s Parsons Brinckarhoff, Singapore.[18] Although NESPAK is a local consultant, the Ring Road detailed feasibility study is supervised by Dr William C. Taylor, a professor at Michigan State University, USA.[19] Under his supervision, a traffic model was developed to account for traffic over the next twenty years. The model is designed for time saving based on a cost-benefit analysis.

Besides the interchange, the Punjab government has also allocated funds for upgrading the Bund Road from Niazi Chowk to Babu Sabu (Phase 1) according to the proposed Ring Road cross-section shown in Figure 7.2. As mentioned before, one section of the Ring Road, from Babu Sabu to Niaz Baig (Lahore Bypass), is a limited access expressway that is already in use to provide a link to the Lahore–Islamabad Motorway.

FIGURE 7.2

CROSS SECTION OF RING ROAD AT BABU SABU TO NIAZI CHOWK (PHASE 1)

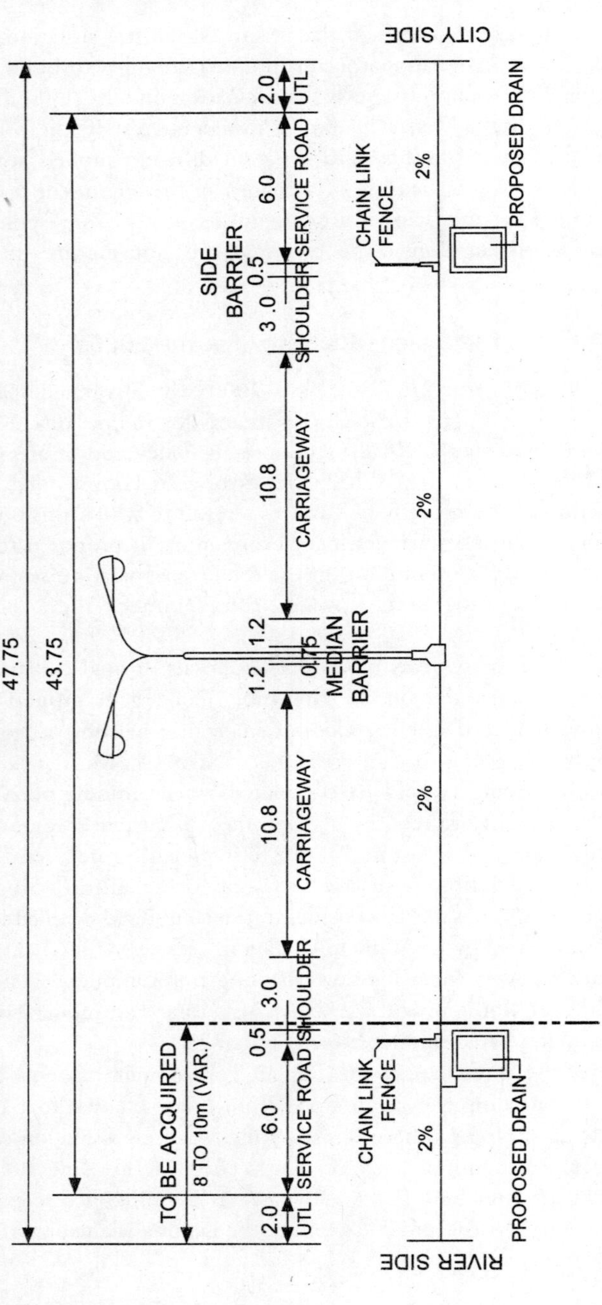

Source: Information provided by NESPAK

Due to lack of financial resources, the Punjab government continues to look for international development institutions to fund completion of the remaining part of the Lahore Ring Road; in pursuit of this objective it made presentations to the Pakistan Development Forum[20] in 2004 and 2005. A brief summary of these efforts is shown in Table 7.1.

It has been established that the Ring Road project was proposed at the time when road development was adopted as the main transport policy paradigm at the national level. It is important to note that after the first proposal for the Ring Road, three comprehensive master plan studies were prepared as well as many smaller studies, but all these enriched the Ring Road proposal rather than raising any questions. To answer the question of why there was no opposition to the project, the discussion will turn to the institutional aspect of project identification, finance and execution.

TABLE 7.1

SUMMARY OF THE EVOLUTION OF THE LAHORE RING ROAD PROJECT

Year	*Report/Project*	**Planning/Financial Allocation/ Construction**
1965	Master Plan for Greater Lahore (1965–80)	* Proposal of Ring Road in the name of 'Circumferential Arterial Road'
1980	Lahore Urban Development and Traffic Study (LUDTS)	* Proposal of Ring Road in the name of 'Southern Bypass'
1980	National Highway Board Proposal	* Stress on one section of Lahore Ring Road, as 'Lahore Bypass'
1988	NESPAK Ring Road Proposal	* Lahore Ring Road Proposal along with financial application (PC1)
1988	Punjab Urban Development Project	* New alignment of Lahore Ring Road
1991	National Highway Authority	* Improvement of some section of Ring Road as a part of National Highway
1992	Comprehensive Study of Transportation System in Lahore (CSTS) (1990–2010)	* 62 km. of Ring Road proposal * Series of other semi Ring Roads
1992	Lahore Ring Road Feasibility Study	* The detailed proposal of 65 km. of Ring Road having 20 km. diameter around the city
1995	TEPA Efforts	* Phase 1 of land acquisition of Ring Road

1997	Opening of Lahore Bypass (one section of Ring Road)	* Lahore Bypass as an part of Lahore-Islamabad Motorway
1999	Lahore Urban Transport Project	* Proposal for the complete construction of Ring Road and allied grade-separated junctions
1999	Ring Road on BOT	* Messer Daewoo's interest to construct Ring Road on BOT
2001	Integrated Master Plan for Lahore (2001–2021)	* Ring Road as a part of 'Road Development Policy and Strategy' * Ring Road in short tem investment programme
2004	The Feasibility of Lahore Ring Road	* Ring Road in different section * NESPAK (Dr William C. Taylor) and M/s Parsons Brinckarhoff, Singapore are developing traffic models for exact alignment
2004	Official Construction of Ring Road commences	* Construction started from interchange at Niazi Chowk * Construction and upgrading on Niazi Chowk to Babu Sabu road Section
Source: Compiled by author		

INSTITUTIONAL PATH DEPENDENCE IN THE LAHORE RING ROAD

Since twenty-nine federal, provincial and local institutions, in addition to international institutions, were involved in the Lahore Ring Road decision-making for over forty years, institutional analysis is an appropriate way of investigating this project. The roles of these institutions in planning, financing, developing and constructing the Lahore Ring Road, as also the operational aspects of the project, need to be explained. The purpose of this section is to explore all the various institutions that were responsible for identifying, financing and executing the Lahore Ring Road project over time. The discussion will be divided into four main sections: (1) planning, (2) financing, (3) execution and (4) land use, and social and environmental approvals. Finally, some issues related to the Ring Road project will be identified.

Project Identification/Planning

a) Role of the International Institutions and their Associated Foreign Consultants

The role of international development institutions in planning and identifying the Lahore Ring Road project was very significant from the beginning. The project was first identified by foreign consultants who prepared the Master Plan for Greater Lahore (1965–80) in the early 1960s. Due to shortage of technical staff in the Lahore Municipal Corporation and the Lahore Improvement Trust, a separate project office was set up in February 1961, and assigned the responsibility for preparing a master plan for Greater Lahore. The master plan project office mainly relied on the advice of Mr P.W.G Powell, Colombo Plan advisor on town planning and housing who was already working as an advisor in the provincial C&W department and Mr A.R. Hye, consulting architect to the West Pakistan government. These two were assisted by three Pakistani counterparts with some town-planning qualifications. The master plan project office published an interim report in 1962 followed by an unapproved master plan in 1965. The plan was formally approved in 1973. It is important to note that the first master plan for Greater Lahore was prepared as a result of the recommendation in the national Second Five-Year Plan (1960–65). The discussion in Chapters 3 and 5 concluded that the involvement of foreign consultants and international development institutions in the preparation of the First and Second Five-Year Plans (1955–65) was on account of weak technical and financial capacities of local institutions. A similar institutional weakness in the provincial and local government organizations resulted in the involvement of foreign consultants for the preparation of the Master Plan for Greater Lahore.

The Lahore Ring Road project was retained in the LUDTS which was prepared mainly by Messrs Halcrow Fox (UK) and the Danish consultants CCH Associated[21] at the request of the LDA. A local consultant, BMK Associates, helped their foreign counterparts in gathering data. The study was financed by credit from the IDA/World Bank, while members of the Bank's mission in Pakistan provided advice and guidance to the study group.[22]

Like LUDTS, the CSTS (1990–2010) had also stressed the need to build the Ring Road albeit with a modified alignment as part of their proposed transport master plan for 2010. CSTS was completed with the technical and financial assistance of JICA in 1991. A team of Japanese transport experts visited Lahore for discussions while actual studies were completed in Japan.[23] Although the TEPA was to collaborate with the JICA in preparing the transport master plan, this was limited to providing local information and collecting necessary data.

The World Bank-financed PUDP (1988–97) presented the Ring Road as an economically feasible project to circulate future traffic within the Lahore Metropolitan Area.[24] It supported this project on the basis of the 'Lahore Ring Road Feasibility Study' completed with the financial support of its ongoing PUDP in 1992. Mott MacDonald International Ltd. (UK) in association with MMP Pakistan (Pvt.) Ltd and Engineering Consultants (Pvt.) Ltd conducted the feasibility study of the Lahore Ring Road in 1992. Later, the consultants W.S. Atkins adjusted one section of the Ring Road into the Lahore–Islamabad Motorway Project which was completed in 1997.

Finally, the Lahore Ring Road was an integral part of the LUTP proposed by the TEPA and the World Bank in 1999. However, for political reasons the World Bank later dropped this project. The TEPA encouraged Messrs Daewoo to build the Ring Road on a BOT basis; the latter engaged Mott MacDonald International to work with different models and various alignments. The Mott MacDonald study, however, did not recommend this project on account of the change in alignment.

In summary, international development institutions, e.g. the World Bank, JICA and associated foreign consultants were fully involved in identifying, elevating and finalizing the details of the Ring Road in Lahore.

b) Role of the Federal Government

The federal government organizations did little to actually initiate proposals for the Lahore Ring Road project but they were instrumental in sourcing foreign aid and consultants for conducting studies that proposed the project. For example, the Master Plan for Greater Lahore (1965–1980) was prepared on recommendations made in the Second Five-Year Plan (1960–65) formulated by the federal Planning Commission with help of the Harvard Group. Similarly, the Planning Commission in collaboration with the Harvard Group arranged for and then engaged consultants for preparing the first master plan for Lahore. Every master plan had the involvement of international institutions and foreign consultants organized by the Planning Commission and each plan supported the Ring Road.

More specifically, in the early 1980s, the NHA was involved in proposing a bypass section of the Ring Road. Subsequently, the NTRC submitted a report on behalf of the NHA to propose a new bridge on the River Ravi, along with the improvement of the westward bound Bund Road known as the Lahore Bypass (one section of the Ring Road) in 1991. In short, federal government organizations have acted as coordinators in the planning of this project.

c) Role of the Provincial Government

Before enactment of the Local Government Ordinance of 2001, organizations of the Punjab government played a central role in the planning, funding, development, operation and regulation of the Ring Road in Lahore. At the planning and proposal stage, the provincial housing and physical planning department supervised preparation of the Master Plan for Greater Lahore (1965–1980) which became the first document to propose the Ring Road project. The other master plans (mainly land-use plans) which emphasized the necessity of the Ring Road project were completed under the LDA. Until 2001, the chief minister of the Punjab was the chairman of the LDA.

More specifically, the Lahore Ring Road was one of the stronger recommendations made in the Integrated Master Plan for Lahore (2001–2021) prepared by NESPAK—the first plan to be entirely prepared by local consultants—with funding from the LDA. Clearly, the Punjab government organizations were directly involved in the planning proposal for the Lahore Ring Road.

d) Role of the Local Government

The local government structure in Lahore earlier comprised of the Lahore Municipal (later Metropolitan) Corporation but its responsibilities were limited to paving streets in inner city areas. As the Ring Road was proposed to be built on the periphery of the city, the Lahore Metropolitan Corporation did not have a direct role at the planning stage; that part of Lahore was under jurisdiction of the Lahore Improvement Trust (later the LDA) with the Punjab chief minister its chairman. As a result of the devolution plan of 2001, the Lahore Metropolitan Corporation was dissolved and the LDA was placed under the LCDG. Recently the LCDG has been given greater powers but it has not played any direct role in the planning of the Ring Road. Finally, there was no public consultation process to elicit views about this project from the local communities, academics and concerned NGOs.

Project Funding and Finance

The aim of this section is to explore the institutional financing of various activities in the Lahore Ring Road project. The discussion is divided into three categories: (1) financing the studies, (2) financing proposed by the studies, and (3) financing construction of the Ring Road.

a) Role of the International Institutions

The international development institutions have played a significant role in financing urban development studies. The Ring Road project was first conceived in the Master Plan for Greater Lahore (1965–1980) in the early 1960s at a time when Pakistan was receiving a substantial amount of aid and soft loans; part of this assistance helped in funding this and various other planning and development studies. In this way, LUDTS (1980–2000) which retained the concept of the Ring Road was financed by credit from the IDA/World Bank.[25] Later, a series of proposals for the Ring Road were developed by CSTS (1990–2010) that was financed by the JICA. Finally, the detailed feasibility study for the Lahore Ring Road was prepared in 1992 with funding from the World Bank that was already available for the ongoing PUDP. Thus, over a period of forty years, international institutions played a central role in funding studies that consistently proposed building of the Lahore Ring Road.

This section will explore the quantum of finance proposed by these studies for the planning and construction of the Ring Road. LUDTS (1980–2000) identified that Lahore faced a serious problem of funding for the development of its infrastructure. Investment in building roads is part of other public sector expenditure and needs to show community benefit before being sanctioned.[26] Therefore, the five-year highway programme (1981–86) proposed by LUDTS (1980) included some sections of the Ring Road that covered the Ring Road from Bypass Road–Babu Sabu to Amar Kot and widening of the Niaz Baig Road. The plan proposed to allocate Rs16.7 million for construction and improvement of these sections from 1981 to 1986.[27]

The CSTS (1990–2010) estimated the construction cost of the Ring Road at Rs272 million in 1990 as shown in Table 7.2.

TABLE 7.2
RING ROAD COST IN COMPREHENSIVE STUDY ON TRANSPORTATION SYSTEM IN LAHORE

	Financial Cost in Rs/ Km	*Total Length (in Km) or (Total Nos. of Flyovers and Bridges)*	*Total Cost (Rs)*
Urban Road (4-lanes)	42,300,000	*53.75*	2,2 73,625,000
Flyovers (above roads)	32,800,000	*8*	262,400,000
Flyovers (above Railway)	41,500,000	*2*	83,000,000
Ravi Bridge	102,100,000	*1*	102,100,000
Total Cost			2,721,125,000
Source: Extracted from TEPA and JICA 1992: Table 8.4.1 and Table 8.4.2, 8-13			

This cost did include money needed for upgrading the existing Bund Road (one section of the Ring Road) and the cost of land acquisition required to secure an appropriate right-of-way (ROW) for the remaining sections of the Ring Road. The CSTS recommended that funds should be allocated to meet the estimated cost of construction of the Ring Road.

One section of the Lahore Ring Road called the Lahore Bypass was completed in 1997 on a BOOT basis by the private sector—a consortium of mainly international companies, named the Lahore Bypass Bridge and Road Company (Pvt.) Ltd.[28]

The World Bank has been the main source of funding for road development in Lahore, financing 79 per cent of expenditure in the PUDP (1988–97). The land acquisition and partial construction of the Ring Road up to 1997 was made possible by World Bank funding. It was anticipated that the Lahore Ring Road would be completed under the LUTP planned to start in 1999. The World Bank was prepared to finance 69 per cent of the project cost with the remaining 31 per cent committed by the Punjab government but eventually the project failed to materialize. Later Messrs Daewoo conducted a feasibility study to build the Ring Road on a BOT basis but this too did not mature.

b) Role of the Federal Government

The federal government organizations had little direct role in financing planning studies or in building the Ring Road in Lahore. For example, the NHA and NTRC prepared proposals for the Ring Road from within their own budgets. However, the federal Planning Commission was directly involved in arranging finance from the international development institutions for preparing master plans and conducting feasibility studies for the Ring Road.

Although the federal government has not been involved in financing the construction of the Ring Road, but considering the financial position of the Punjab government (P&D and C&W departments), this road cannot be built without the assistance of the federal government or international institutions.

c) Role of the Provincial Government

The Punjab government has over time changed its role in financing the Lahore Ring Road project. As described earlier, most master plan studies were financed entirely or partially by aid or credit from international institutions. In cases of partial funding the remaining components of the total costs were financed by the Punjab government from its ADP prepared by the P&D department. However, the current Integrated Master Plan for Lahore (2001–21) was fully financed by the LDA, recalling that the Punjab chief minister was the LDA chairman. The plan recommended building the entire Ring Road as early as possible. As a result, 90 per cent (Rs16,000 million) of the budget for the short-term investment programme was proposed to be allocated for building the Ring Road.[29]

During the 1990s, the TEPA and the LDA, with the Punjab chief minister its chairman, were the agencies that were planning and building the Lahore Ring Road. After completing feasibility studies, land acquisition and partial construction of the Ring Road was started by the TEPA. The World Bank and the Punjab government funded the work until 1997. Details of activities and the money estimated or allocated (as the case might be) are shown in Table 7.3.

Table 7.3 highlights that after completion of the PUDP in 1997, the Punjab government was the main source of funding, but after the Local Government Ordinance 2001 was promulgated, the TEPA was detached from the LDA and placed under control of the LCDG. In view of the new institutional arrangement, the Punjab government transferred the project to the C&W department in 2002. As the office that had primary responsibility for the construction of the Lahore Ring Road this department was the sponsoring agency for the construction of the first interchange of the Ring Road at Niazi Chowk. The PC1 estimated a cost of Rs720.9 million for the construction of the interchange[30] but the Punjab government approved Rs939.6 million[31] for the project in its ADP for 2004–5 (ADP No. 287). From this total amount Rs450 million was to be spent during 2004–5.[32] It was expected that the interchange would be completed in mid-2006. Besides the Niazi Chowk interchange, the ADP[33] for 2004–5 also approved Rs779.334 million for the construction of one section of the Ring Road from Niazi Chowk to Saggian Chowk.

TABLE 7.3
TEPA Budget Estimates and Receipts for the Lahore Ring Road

Budget Estimate Years	*Ring Road Development Activity*	*Money Estimated/ Expenditure/ Allocated in Rs (million)*	*Funding Authority*
1993–4	Land acquisition and partial construction of the Ring Road	24.15	Govt. of Punjab (21%) and the World Bank (79%)
1994–5	Land acquisition and partial construction of the Ring Road	30.65	As above
1995–6	Land acquisition and partial construction of the Ring Road	24.05	As above
1996–7	Land acquisition and partial construction of the Ring Road	13.65	As above
1997–8	Land acquisition and partial construction of the Ring Road * Consultancy for the Ring Road (Rs10 million) was not approved by the Govt. of Pakistan	5.34	Govt. of Punjab
1998–9	Land acquisition and partial construction of the Ring Road	0.87	Govt. of Punjab

1999–2000	Land acquisition and partial construction of the Ring Road	Data is not available	Data is not available
2000–1	Land acquisition and partial construction of the Ring Road	17.57	Govt. of Punjab
2001–2	Land acquisition and partial construction of the Ring Road	15.85	Govt. of Punjab
2002–3	Construction of the Ring Road	15.85	Govt. of Punjab
2003–4	Construction of the Ring Road	15.85	Govt. of Punjab
2004–5*	Construction of the Ring Road	15.85	Govt. of Punjab

Source: Figures taken from the TEPA budget estimates of several years
* The C&W department took over the Lahore Ring Road project from the TEPA. The Punjab government allocated Rs650 million for the construction of the interchange at Niazi Chowk and one portion of the Ring Road from Niazi Chowk to Saggian Chowk.

Rs200 million was allocated for the construction of this section during 2004–5 which did not include the cost of land acquisition and resettlement that would be necessary for building this section. The Punjab government estimated an allocation of Rs20,000 million for the Ring Road for 2005 and 2006 but actually allocated Rs18,000 million (in the name of Special Infrastructure Programme) in their development budget for 2006 and 2007. This financial allocation covered 18 per cent of total development budget for the entire province during 2006–7.[34] The overall cost of construction of the Ring Road has not yet been declared due to pending alignment and land acquisition issues but the unit cost of the Lahore Ring Road was estimated as: (1) at grade alignment Rs100 million per kilometre, (2) elevated alignment Rs1000 million per kilometre, (3) interchange (average) Rs500 million for each, (4) flyover (average) Rs100 million for each, and (5) pedestrian overhead bridges (average) Rs6 million for each.[35] In 1998, the TEPA estimated the cost of construction of the Lahore Road at Rs16.9 billion.[36] To summarize this discussion, one can point to evidence that the Punjab government has been one of the prime sources of funding through its ADP for recent construction activities of the Lahore Ring Road project.

d) Role of the Local Government

The local government in Lahore has played no role in financing the Lahore Ring Road project because of the weak financial position of the Lahore Metropolitan Corporation.

PROJECT CONSTRUCTION PROGRESS/EXECUTION/PROCUREMENT/ OPERATION

Notwithstanding the many proposals for the Ring Road, actual progress on the project was slow (see Fig. 7.3). The Master Plan for Greater Lahore

FIGURE 7.3

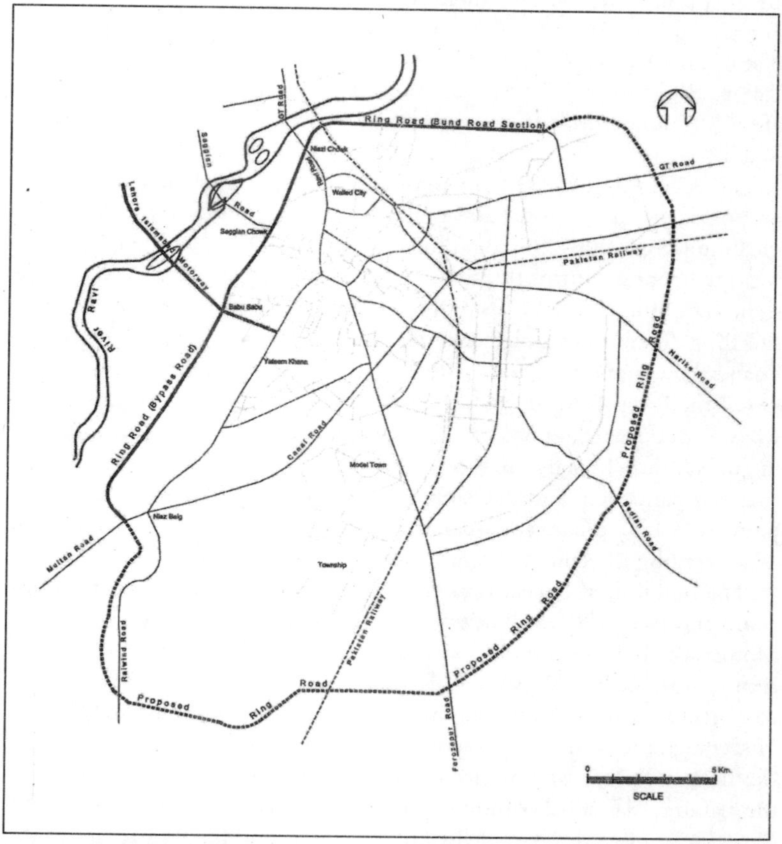

Source: LDA 2004 a&b.

(1965–80) proposed that acquisition of right-of-way (ROW) for the proposed circumferential road (current Ring Road) and the inner Ring Road should be made possible during the first phase of the Plan (1965–70). It was further proposed that construction of the proposed circumferential road should start in the second phase of the Plan (1970–5).[37] However, no progress was made in implementing the recommendations to construct the Ring Road.

LUDTS (1980–2000) later proposed a five-year investment programme from 1981 to 1986 which included schemes for the construction of new roads and improvement of the existing ones. It was proposed that land should be acquired to reserve the right-of-way for major roads including the Ring Road after the preparation of detailed physical surveys, accurate air mapping and careful alignment within the physical framework provided by the Study.[38] The five-year investment plan (1981–6) proposed that 35.9 per cent of the total budget for transport be allocated for the Lahore Ring Road.[39] Before the next transport study in 1992, an ROW of 12.50 kilometres had been acquired for various sections of the Ring Road.[40]

The CSTS (1990–2010) recommended that implementation of its proposals be spread over three phases, i.e., the short term (1992–5), medium term (1996–2000) and the long term (2001–10). The Ring Road was included as a priority project for a detailed feasibility study in the short-term phase. The CSTS proposed that a detailed feasibility study for the Ring Road be completed as part of the ongoing PUDP (1988–97) financed by the World Bank.[41] Moreover, the study pointed out that 12.50 km of the Ring Road ROW was available for construction to commence, while a stretch of 9 km had to be acquired to complete the priority section in the south.[42] In light of these observations, the CSTS recommended that part construction of the Ring Road should be completed in the short-term (1992–5) phase and that the remaining construction should be completed by the end of the medium term (1996–2000).[43]

The building of one section of the Ring Road—the Lahore Bypass—commenced in 1991, to link the city of Lahore to the Lahore–Islamabad Motorway. This section was made on a BOOT basis by the private sector—the Lahore Bypass Bridge and Road Company (Pvt.) Ltd, a consortium of mainly international companies.[44] A new section between Niaz Baig and Babu Sabu was built along the existing *bund*. Moreover, the Lahore Bypass was connected to the 'zero point' (starting point) of the Lahore–Islamabad Motorway by making a new bridge on the River Ravi. The project also remodelled the existing Bund Road (existing section of the Ring Road)—mainly dualization of the road from Babu Sabu to Niazi Chowk (7.9 km) according to the design criteria and approximate cost described in Box 7.1. Due to the existing land-use and activity pattern, limited access was not possible in the existing section as proposed in the feasibility studies. A flyover at Niazi Chowk near the River Ravi was proposed which is now under construction as part of the Ring Road Project.[45]

BOX 7.1

Lahore Ring Road—Design Criteria

- Design Speed—100 km/hr
- Lane Width 3.6 m
- No. of Lanes—3 Dual Carriageway
- Outer Shoulder—3.0 m
- Inner Shoulder—1.2 m
- Service Roads—in built up area
- Limited Access Urban Expressway

Lahore Ring Road—Estimated Unit Cost

- At Grade Alignment—Rs100 million per km
- Elevated Alignment—Rs1000 million per km
- Interchange (average)—Rs500 million each
- Flyover for Lateral Roads (average)—Rs100 million each
- Pedestrian Overhead Bridge (average)—Rs6 million each
- Cost of existing radial roads improvement, relocation of utilities, land acquisition and compensations is not included

Source: Information provided by NESPAK as presentation slides

In the early 1990s, the TEPA realized that the Lahore Bypass section of the Ring Road would mainly serve intercity traffic rather than the metropolitan area. To meet the needs of transportation for the metropolitan area, the TEPA carried out a detailed feasibility study of the Ring Road in 1991 along with the proposed construction of the bypass section. The detailed design of 18 km of Phase 1, from Yateem Khana interchange to Ferozepur Road, was also completed as part of the feasibility study of the entire Ring Road project. In light of the Lahore Ring Road feasibility study in 1992, the TEPA started acquisition of land and secured a stretch of around 9 km as part of Phase 1 in 1995. Moreover, the TEPA built 6.5 km of one section as a single carriageway

(Chief Engineer, TEPA), while one section of the Lahore Ring Road (Bypass) was opened for traffic in 1997 as part of the Lahore–Islamabad Motorway.

In 1999, the Punjab government tried to complete the Ring Road by initiating the LUTP mainly funded by the World Bank. Later Messrs Daewoo was also involved in efforts to construct the Ring Road on a BOT basis. However, neither of the two initiatives could be finalized.

In light of the Integrated Master Plan for Lahore (2001–21), actual construction of the Ring Road started on 22 November 2004 with a ground-breaking ceremony attended by the President of Pakistan. Soon thereafter, the construction of the interchange at Niazi Chowk, an important junction of the Ring Road, commenced. Currently the existing Bund Road, an important section of the project, is being remodelled to bring it at par with standard of the Ring Road (limited access expressway).

The local government in Lahore is not involved in the construction of the Ring Road because a large segment of the road lies outside the jurisdiction of the Lahore Metropolitan Corporation. Although the LDA and TEPA were involved in proposing the project and in its initial construction, this was because the chief minister of the Punjab was then the chairman of the LDA. However, ever since the enactment of the Local Government Ordinance 2001, the LDA is headed by the elected *Nazim* (Lord Mayor) of Lahore. The responsibility for the construction of the Ring Road was therefore transferred from the LDA and TEPA to the C&W department of the Punjab government.

Land Use, Social and Environmental Approvals

The LCDG is responsible for land-use planning, and addressing the social and environmental impact of the Ring Road in Lahore. The LDA as part of the LCDG regulates land-use according to its master plan; the Integrated Master Plan for 2021 is presently under implementation. The LDA reserved land for building sections of the Ring Road that pass through various housing schemes. The Bund Road section of the Ring Road passes through built-up areas and is expected to displace around 500,000 people and their homes.[46] This prompted the affected residents to file a petition in the Lahore High Court challenging the right of the authorities to build that section of the road. The High Court dismissed the plea on the grounds that the court did not have the right to interfere with the government's policy[47] and besides the project was in the larger public interest.[48]

Similarly, there is a fear that 250,000 people living in 50,000 units in the eastern sections of the Ring Road would be displaced.[49] The Punjab government is providing compensation to the displaced people but residents claim that the price being offered is Rs20,000 per *marla*[50] against a market rate which is about Rs250,000 per *marla*.

In terms of the environment, responsibility for mitigating the environmental impact of the Ring Road lies with the district environment officer, LCDG and the environment protection department of the Punjab government. According to the Pakistan Environmental Protection Act 1997, all major road projects are required to submit EIA reports to the environment department of the province. However, no formal EIA was carried out for the Ring Road project; instead a short environment report was submitted to fulfil the requirement of Part D of the PC1 sent to the P&D department. Nadeem and Hameed found that in Pakistan there is no practice of carrying out EIA studies for public sector projects.[51] There might a few exceptions but even these EIA studies are generally based on inadequate baseline data.[52]

Issues

a) Land Acquisition

The Ring Road is currently being built on existing sections of the road to upgrade these to the standard of a limited access expressway. These sections do not require acquisition of large tracts of land but the entire land needed for building new sections on the southern and eastern sides needs to be acquired. The TEPA secured some part of the required land on the southern side of the Ring Road route as a quid pro quo for according approval to private housing schemes in the area. In addition, the TEPA acquired some more land for the Ring Road on the same section with the financial help of the World Bank and the Punjab government in the mid-1990s. But overall, a major chunk of the land has yet to be acquired and this ought to be done soon if the Ring Road is to be completed on time. It is important to mention that most of this land is under private ownership but under the provisions of the Punjab Land Acquisition Act, the government can acquire privately owned land for public purposes by buying it at the market price. The price and the time within which this land is acquired remains a major issue for the successful completion of the Ring Road.

b) Funding

The construction and possible operation and maintenance of the entire Ring Road will require a dedicated, long term commitment and allocation of large amounts of funds. International development institutions provided the initial finance for conducting feasibility studies, while the early construction and acquisition of land was financed by the World Bank and the Punjab government. The latter continues to finance the construction but it lacks the financial capacity to fund the entire cost of the project (in excess of Rs16.9 billion). It must therefore look for financial assistance from the federal government and international development institutions to meet the shortfall.

c) Construction

At present the C&W department of the Punjab government is responsible for the construction of the Ring Road. However, under the new local government system, the TEPA, as a component of the LCDG, is responsible for the planning, construction and maintenance of roads in Lahore. Therefore, one may expect a change in the responsibilities and role of both provincial and local organizations in order to resolve the jurisdictional dichotomies in the Ring Road project.

d) Operation

The Lahore Ring Road is being made as a limited access expressway designed for a traffic speed of 100 km/hr. The planning documents do not define the operation and maintenance process for the road once it has been completed and commissioned. But taking a cue from the recent past, all new highways in Pakistan have been made toll highways so one may expect the same for the Lahore Ring Road. The Punjab government and the LCDG have no previous experience in managing toll highways because it is the NHA that is responsible for the operation and maintenance of all existing highways in Pakistan. It appears that in the end there might be disagreement amongst the federal, provincial and local governments as each would vie to assume ownership of the Ring Road operation.

e) Environmental and Social Issues

The environmental and social concerns arising due to the planning and construction of the Ring Road were not properly addressed during the decision-making process. The project was continuously opposed by local communities living along the road alignment mainly because their businesses and homes were to be bulldozed without an appropriate resettlement plan. A recent appeal in the court against the alignment of the Ring Road by local residents as stated earlier is one of the examples of these concerns. It is also possible that affected people in collaboration with academics, civil rights activists and NGOs might raise these issues at international fora as they did for the Lyari expressway in Karachi.[53] Since environmental and social institutions are becoming progressively more assertive, it is possible that their activism might impact the ability to complete the project on schedule and within budget.

TECHNICAL PATH DEPENDENCE IN THE LAHORE RING ROAD

'Predict and provide' has been the primary transport planning approach employed in the UK from the last many decades to justify road infrastructure projects.[54] In the 'predict and provide' approach, 'the demand for travel by various modes is extrapolated and then attempts are made to match the supply of infrastructure to that potential demand'.[55] Dimitriou called this approach the 'Urban Transport Planning Process' (UTPP).[56] Within a defined theoretical framework, this study is interested in finding how various approaches were employed to justify the Lahore Ring Road project. Two main approaches, described in the following sections, were adopted over time.

Planning Approach

The proposal for the Ring Road in Lahore, initiated by the Master Plan for Greater Lahore (1965–80) was influenced initially by 'planning thoughts'. Lahore was originally conceived as a 'Garden City',[57] with modern planners visualizing it as 'having suburban, low-density development, green belts and new orbital and peripheral roads'.[58] The central justification for the Ring Road was that it would provide a circumferential link to connect all the main roads of Lahore that were radial in nature. The circumferential design of the Ring Road was planned to reduce traffic on Lahore's radial roads and at the same time connect its satellite towns to the city.[59] The master plan believed that the proposed Ring Road would reduce tendencies for ribbon development along radial highways.[60]

The LUDTS (1980–2000) proposed the Ring Road project based on the planning concept that it would help manage the south and south-westward growth of Lahore over the next twenty years.[61] Consequently, the Ring Road or 'Southern Bypass Road' was an integral part of the structure plan rather than the traffic study report. The LUDTS enriched the proposal by postulating an 'area of opportunity'—a planned corridor for employment along the Ring Road.[62]

The documents prepared in the early 1990s e.g. the Lahore Bypass study, CSTS (1990–2010) and Lahore Ring Road Feasibility Study also justified the Ring Road on the basis of this planning concept. The main justification of the Lahore Bypass proposal (one section of the Ring Road) was to accommodate increasing population and traffic in Lahore.[63] The study calculated that the population and motorized traffic of Lahore were growing at a rate of 3.8 and 6.1 per cent per year respectively. These figures were used to forecast future population (12 million) and traffic by 2015. In view of these trends it was postulated that large urban areas would be required to accommodate the future population of Lahore and identified the southern and western sides of the city as potential growth areas. The Lahore Bypass was therefore proposed to connect through traffic to the Lahore central area. The Lahore Ring Road Feasibility Study also justified the Ring Road as an integral component of strategic planning. The study pointed out that the Grand Trunk Road (GT Road) was the principal link connecting northern Pakistan to its southern part, right up to Karachi. It also observed that major commercial and industrial activities were established along the GT Road with 80 per cent of the population living in its vicinity. Lahore presently sits astride this route and it was therefore argued that building the Ring Road would ensure minimal disruption of all north-south traffic on the GT Road. Also, due to the strategic nature of the GT Road, it was proposed that the standard and design of the Ring Road should be such that it would permit fast movement of traffic without potential hazards like pedestrian access.[64] In short, the Lahore Ring Road was first justified as being part of a planning strategy to accommodate the future growth of Lahore and a crucial link for the free flow of traffic on the strategically important GT Road.

Predict and Provide/Urban Transport Planning Process

The Master Plan for Greater Lahore (1965–1980) and the LUDTS (1980–2000) used the 'predict and provide' approach as a standard technique that forecast future traffic to justify the Ring Road proposal. As described in Chapter 5, the 'predict and provide' approach in the

Pakistani context is different from that in developed countries where the historical growth rate of traffic is used as a reference to establish trends and project future traffic volumes. In contrast, both studies for the Ring Road projected future traffic by not just taking into account the volume of current traffic and its growth rate in Lahore, but by also considering issues like land-use development, population growth and growth of per capita GNP.[65] The LUDTS (1980–2000) also considered other factors such as the composition of traffic, future traffic speed, strategic needs, accident records, overall road network capacities, existing local standards and the cost of implementation of road infrastructure proposals.[66] Thus, a combination of these factors and land-use development were made the bases for justifying the Ring Road project in the 1960s and early 1980s. It will be useful to examine each of these criteria separately.

The projected growth of per capita GNP was one of the factors that were used to justify the project; this was rooted in the firm belief that increased economic activity would create a low-density urban sprawl and a high volume of private vehicles.[67] It was therefore argued that an increase in the per capita GNP would change the growth rate and composition of traffic in Lahore. These assumptions were however in contradiction to what was observed in the LUDTS in the early 1980s. For example, the LDA and the World Bank found that:

> ...levels of per capita income and car ownership are very low in Lahore as compared to the situation in other rapidly growing urban centres throughout the world. This fact is reflected in the importance of walking, cycling and intermediate forms of public transport as means of moving about the city.[68]

The document also stated that:

> ...the problems that beset western cities are not to be found in Lahore. There is little congestion, low private ownership, an expanding demand for public transport, low wage rates, no noticeable traffic peaks and high residential densities. The city is well adapted to use of the bicycle, which is patronized to a degree that would be the envy of many city planners elsewhere. It has also developed a good, economical system of rickshaw taxis and in certain areas *tongas* fulfil a need and provide environmentally admirable services.[69]

These remarks underline the fact that the per capita income in Lahore was low, and that transport characteristics of the city in comparison to other cities of the world were quite different.

In terms of speed, the LUDTS noted that:

> ...observed speeds throughout Lahore can in general be described as satisfactory. They do not indicate that the level of congestion is severe, as over most of the network free-flow conditions exist. Car speeds over almost the entire network away from central area rarely dropped below 20 km/hr[70]... observed travel speeds on the network reflect this overall lack of congestion[71]... little congestion is evident and journey speeds are generally satisfactory for an urban area [in Lahore][72]...primary roads are capable of offering very quick travel times.[73]

These observations on the travel speed in Lahore reflect a situation quite different to one that could justify the urgent need to build a new high-speed road.

As regards the road network condition in Lahore, LUDTS found that:

> Lahore has the basis of a good highway network which offers very generous spatial provision[74]...the main road network [of Lahore] is well developed and capable of accommodating existing and future flows. Flow forecast for 1986 are expected to be accommodated comfortably without major network changes[75]...over half the traffic entering the central area arrives by seven main routes and these are mostly multi-lane carriageways well able to accommodate current and forecast traffic flows (1986)[76]...road widths currently available are generous in relation to the volumes of traffic handled[77]...Lahore's road infrastructure is basically adequate; the hierarchy of primary, secondary and local roads has the potential to meet traffic demands now and in the next five-years[78]...lack of highway infrastructure, ineffective and under-financed institutions and pronounced peaks in traffic flow caused by high car-ownership levels are typical causes of road traffic problems found in many other cities. This is not the case in Lahore where traffic congestion is found in only a few areas and traffic running speeds are acceptable over most of the network.[79]

Clearly, the road network in Lahore was more than adequate after two decades when the Ring Road proposal was first initiated.

The LUDTS had also stressed that local standards should be adopted in proposing transport infrastructure projects. When the current study applied this criterion to the Ring Road project, it noted that the Master Plan for Greater Lahore (1965–80) had proposed that the Ring Road be made as an arterial road in view of the projected nature and volume of traffic. The Plan proposed (in the early 1960s) that the Ring Road should be built as an 8-lane, limited access divided highway with complete grade-separation and a minimum ROW of 250 ft. It was further suggested that service lanes should be provided within the ROW of the arterial road for giving access to abutting properties.[80] A road design of this type obviously did not take into consideration the characteristics and standard of the local traffic. Moreover, the cost of implementing the Ring Road project with these specifications was beyond the financial capacity of the local institutions. The LUDTS noted that 'Lahore will continue to have serious problems of financing infrastructure investment. The road will have to compete with all other sectors of public expenditure and as a result all proposed road investment should show some kind of net community return before being sanctioned'.[81] In addition, the discussion in Chapters 1 and 6 demonstrated that Lahore has a high rate of road accidents as a result of the encouragement given to the growth of motor vehicles. The provision of more road infrastructure must add to these accidents in the future. In summary, LUDTS (1980–2000) did not make a strong case for justifying the Ring Road project.

The CSTS (1990–2010) employed the UTPP for predicting future traffic and made that the basis for new transport infrastructure proposals for Lahore. The UTPP was conducted with a four-stage model, i.e., trip generation, trip distribution, modal split and traffic assignment to predict traffic for the year 2010. The UTPP identified and quantified 'capacity deficiency' in the road network, calculating the deficiency by predicting 'average travel speed' and 'average volume to capacity (V/C) ratio'.[82] Based on these calculations, it was predicted that travel speed on the road network would decrease from 28 km/hr in 1990 to 11 km/hr in 2010 due to increase in the volume of traffic. Similarly, the average volume to capacity ratio of the road network would increase from 0.9 to 1.7 in 2010.[83] On the basis of these analyses new roads were proposed in the CSTS.

The CSTS estimated new road requirements in the southern section of the Ring Road by quantifying traffic passing over established screen-line stations. These surveys identified that the density of the arterial road network is relatively low in the southern suburbs, the area proposed for future urban growth. Moreover, the CSTS pointed out that Lahore had a radial arterial road network but had a serious deficiency of circumferential roads.[84] In light of these analyses and arguments, the CSTS developed a conceptual network pattern as a planning policy which included several radial Ring Roads to accommodate future traffic demand.[85] The cross-section of the Ring Road proposed by the CSTS has 4-lanes with flyovers at major intersections.

The Lahore Ring Road Feasibility Study in 1992 also employed a four-stage model, i.e., trip generation, trip distribution, modal split and traffic assignment as a part of the UTPP to justify the Ring Road proposal.[86] The study used traffic counts, and household and roadside interview data gathered during the CSTS. However, data such as traffic volume, speed and composition on major roads in southern Lahore was collected through primary sources. These data were used for developing a 'person trip matrix' though traffic modelling; the matrixes were then converted to a 'vehicle trip matrix' in order to provide accurate figures for vehicle numbers. However, the composition of vehicles was adjusted by the 'passenger car unit (PCU) matrix' which did not take into consideration buses, mini-buses and non-motorized transport. The reason for neglecting public transport in the PCU matrix was mainly the assumption that the proportion of such vehicles in future Ring Road traffic would be negligible. In addition, population growth and anticipated per capita GNP were also considered for projecting future traffic.[87]

The priority section of the Ring Road was to be completed by 1996 with the entire road set for completion by 2015. It was expected that the commissioning of the Ring Road would bring about a significant reduction in the traffic flow in different parts of Lahore.[88] It was argued that the main role of the Ring Road was to provide a circumferential access for dispersion of traffic and thereby relieve congestion on the radial roads.[89] On the basis of future traffic data and emphasis on design speed of 100 km/hr, the Ring Road was designed to be a limited access, 6-lane highway[90] with service lanes, roundabouts, and priority and grade-separated junctions.

The PC1 of Niazi Chowk on the Ring Road justified the work by projecting traffic volumes (only motorized vehicles) from 2004 to 2024, which was calculated on the basis of the growth rate of three factors: (1) traffic over the past years, (2) fuel consumption by the transport sector, and (3) the GDP.[91] Traffic speed was used as a prime factor for designing the lane requirement at Niazi Chowk, i.e., an average speed of 90 km/hr was used to justify a 6-lane dual carriageway.[92]

To sum up, it has been shown that the Lahore Ring Road was first initiated as a planning proposal but was later justified on the basis of 'predict and provide' or the UTPP technique for predicting future traffic. This must be seen in the context of Dimitriou's finding that there is a similarity in the principal function of this technique in both developing countries and industrialized countries, i.e., to justify transport infrastructure investment decisions.[93] The central assumption of the UTPP is that people are willing to spend more money on new road infrastructure to reduce travel time by alleviating congestion and increasing speed. This assumption may be true for developed countries but aside from perhaps a handful of commuters a vast majority of people in developing countries would opt for a longer travel time and prefer to pay less.[94] Besides this, the UTPP does not take into account non-motorized transport, the need to improve public transport, and the social and cultural differences between the developed and developing countries.[95] This discussion affirms the argument that the UTPP was the principal transport technique used for justifying the Ring Road project, which is similar to practices elsewhere in the world for planning transport investment. The Lahore Ring Road project demonstrates its technical path dependence, based as it is on the conventional planning concept (Garden City) developed in Western countries and the UTPP technique that is used in similar ways across the world.

DISCURSIVE PATH DEPENDENCE IN THE LAHORE RING ROAD

The Lahore Ring Road is one of the examples in which a culture of road building is promoted by a discursive knot of storylines supported and publicized by national and international institutions. The Ring Road was not proposed to serve an existing, low-density urban area as is the practice in the USA but was planned to generate urban development in the periphery of the city as was done in Melbourne, Australia.[96] Using the theoretical framework developed here, the discourse field will be examined to determine if transport policies became dependent upon specific ideas and arguments that were in the minds of decision-makers. The purpose of this section is to report the storylines that formed the discourse on the Lahore Ring Road project. To this end, I surveyed all proposals for this project starting from the early 1960s during field work conducted in 2004. Besides this, assessments were made of oppositional stories published in newspapers and interviews were conducted with decision-makers, academic and common people to evaluate the influence of these storylines. It emerged that over time the themes of these storylines ranged from 'planning' and 'engineering' to 'economic' and 'environmental and social'.

THE TOWN PLANNER'S TALE

In the early 1960s, international planners and their local counterparts perceived Lahore as a future Garden City—with low density suburban development, a green belt, and orbital and peripheral roads.[97] Based on this concept, the famous storyline in Lahore's urban planning documents is that the future low-density areas south of Lahore can only be served efficiently by a 'circumferential' or 'ring' road'. In short, future population in the outer, low-density suburbs deserve good transport which can be delivered by building [better] roads. Consequently, an 8-lane, limited access Ring Road with complete grade-separated junctions was proposed in the early 1960s.[98]

Similar planning storylines were found during field work when a planner stated that the main aim of the Ring Road was 'to serve population already grown in the peripheral areas of Lahore and to facilitate their private vehicles to deviate unnecessary movement required to go in the city'.[99] As planners believed that the Ring Road would serve low-density, outer suburbs where it would be difficult to provide an efficient public transport system, planning for the Ring Road did not take into consideration public transport trips.

All the urban planning documents prepared during the last forty years proposed the continuous, outward growth of a low-density city in the south and south-west of Lahore, and argued that a grid of roads would best serve this form of growth. These arguments were the basis for justifying the Ring Road, and later, several full and partial rings around the urban area.

Since planners are generally concerned with population growth and housing, the initial argument in proposing the Lahore Ring Road was that by providing circumferential access, urban traffic would remain on arterial roads and out of the newly developed residential areas. Later, the Ring Road was also presented as a strategic link connecting northern Pakistan to its southern part, all the way to the country's main port of Karachi. Since the presence of large population centres and major commercial and industrial activities along this link could retard the free flow of traffic, it was noted that this section would provide through-traffic the possibility to traverse the GT Road with the least disruption. It was therefore proposed that the road should be designed in accordance with the strategic requirements of a primary distributor such as fast traffic speed and no pedestrian access.[100] Low et al.[101] found a similar perception justifying expenditure on arterial roads or bypasses in the English report of Colin Buchanan[102] and of Melbourne Metropolitan Board of Works.[103]

Opponents of these storylines claim that the Ring Road would benefit only a small group of people—less than 20 per cent of Lahore's population who live in the vicinity and have cars[104] or those who live in nearby cities[105]—rather than the majority of people living in Lahore. However, this opposition is only against the route or alignment of the Ring Road rather on the reduction of car use or on the basic issue of building the Ring Road. The route of the Ring Road has been changed several times due to unavailability of cheap land. One planner had this to say: 'The Ring Road was proposed at the right time. But we had not fulfilled the pre-requisite step of acquiring the land on which the road was to be built. If that had been done, we could have built the Ring Road much faster and at a lesser cost'.[106] Although the Punjab government did allocate resources for the project, mainly for land acquisition[107] and construction,[108] these were not enough to cover the cost of building the entire Ring Road. Some planners were of the opinion that the Ring Road project was the outcome of a transport planning approach which focused only on engineering solutions, i.e., building the Ring Road, rather than land-use planning solutions.[109]

There is both truth and falsehood in the claims made in planning reports justifying the Ring Road. For example, even after the analysis of the LDA and World Bank[110] that the 'level of per capita income and car ownership are very low in Lahore as compared to the situation in other rapidly growing urban centres throughout the world', the arguments in favour of the Ring Road did not cease but instead the focus moved to engineering and economic storylines.

The Engineer's Tale

'Predict and provide' is not only the standard method for justifying transport investment decisions as described in the last section but Low et al. argued that it is the model storyline around the world.[111] This storyline is supported by projecting trends of different transport modes and accommodating these trends into the proposed transport system. For Lahore, however, factors such as land-use development, growth of population and growth of pre capita GNP were also considered alongside traffic growth rate to support the Ring Road proposal. Nevertheless, the aim was to accommodate future traffic. For example the chief minister of the Punjab stated: '...one million registered vehicles were plying on the roads in Lahore and the number was increasing every year. Ring Road will help to accommodate these vehicles.'[112]

A similar view was held by a local expert, '...looking at the trend of traffic, the Ring Road is necessary for Lahore'.[113] Perhaps, technical experts provide storylines to support political statements.

Over time, the 'predict and provide' technique started to be modified into a 'missing links' storyline. For example, a senior local expert (Chief Engineer TEPA) stated that, "...the Ring Road will work as 'missing link' between radial routes...data shows that 60 per cent of Lahore's traffic is on the radial roads. Without routes that can act as distributaries, there is a traffic imbalance. So the Ring Road will work as a missing link to provide smooth distribution of traffic'. On similar lines, another expert believed that, '...Lahore needs the Ring Road. It will provide a bypass for the traffic of Lahore'.[114]

The Lahore Ring Road proposal was sustained by two other storylines, 'elimination of congestion and increasing speed of traffic'. 'Congestion' means '*abnormal* accumulation especially of traffic'; this would then suggest that to establish a *normal* condition it would be necessary to eliminate congestion by improving the free flow of road traffic.[115] The TEPA studied the impact of the Ring Road by analysing three aspects: minimum traffic conditions, impact of the priority section and impact of the complete Ring Road.[116] The main finding of the analysis was that there would be a 'reduction of traffic flow' in different ranges. The TEPA study concluded that the Ring Road would alleviate congestion which would be equally beneficial for the public transport.[117] However, other studies found that bus traffic was likely to be very small in volume and did not require to be accommodated in the traffic model developed for the Ring Road. Similar storylines were given by senior experts during interviews who believed that '...a significant difference in the traffic situation would be seen after the completion of the Lahore Ring Road. The width of the road would provide more capacity for vehicles and relieve congestion'.[118] Storylines that 'the Ring Road addresses frequent congestion problems in the metropolis by reducing travel time' can also be found in ADP 2006–7.[119] Low described this as the 'free movement storyline' in his study identifying discursive barriers to sustainable transport in Australian cities.[120] He explained how medical analogies, such as 'heart', 'arteries' and 'bypass' were used in Melbourne's City Link motorway project advertisement to promote this storyline.

The 'elimination of congestion' was followed by the 'increasing speed of traffic' storyline. Medical analogies were used by some experts to promote this, saying that:

> ...roads are like veins in the human body and traffic is like blood that must flow all the time. The speed of vehicle flow is the main factor in urban transport planning. Speed can be maintained by constructing expressways and

> peripheral roads like the Ring Road project in Lahore. These roads can accommodate the increased growth of motorization that has been occurring over the last few decades.[121]

The 'increasing speed' storyline was also noted from the interview of provincial experts, one of whom said that the '...recent widening of roads, building underpasses and the construction of the Ring Road in Lahore will significantly improve the speed of traffic'.[122] All these supportive storylines provide strength to the main storyline of 'predict and provide' and are used to justify the Ring Road proposal.

The Economist's Tale

The first economic storyline runs around 'economic growth needs the Ring Road', similar to the national transport storylines described in Chapter 5. Therefore, future GDP/GNP growth was considered to be an important factor in justifying the Ring Road. This storyline was constructed in the belief that economic growth would result in an increase in the number of private cars, and to accommodate this increase investment in road infrastructure would be necessary.

Over time, the above storyline was inverted to say that 'the Ring Road would bring economic growth'. This storyline was developed and then articulated by the Punjab chief minister in an address on 'Punjab Vision 2020':

> ...[I had observed that] access to communications had a positive effect on the economy of a local area by increasing mobility, reducing transportation time and cost, enhancing access to goods and services and stimulating commercial and other income generating activities through better access to markets.[123]

More specifically concerning the Lahore Ring Road, the Punjab chief minister stated that, the 'Lahore Ring Road will ensure a smooth flow of traffic, help in the development of infrastructure, strengthen the provincial economy and raise the quality of life'.[124] These storylines were projected by an aggressive campaign in the print and electronic media stating that:

> ...the construction of the Ring Road would help to develop industrial zones and will also strengthen the provincial economy...the project would play a vital role in infrastructure development and strengthening the economy...free flow of traffic from the Ring Road will bring commercial and industrial activities in Lahore.[125]

These storylines started to spread out and later emerged in other segments of society as well. For example, in response to the legal case against the Lahore Ring Road, the Lahore High Court declared that 'the Lahore Ring Road was a public welfare project, which would provide better transport and economic facilities to the commuters in the city'. The court remarked that 'even if there were any changes in the original plan, they were presumed to be taken in the interest of the public at large'.[126] The 'roads bring economic growth' storyline was identical to that found by Low et al. in 'Roads to the Future' in Australia[127] and 'Local Economic Competitiveness' by Vigar in the UK.[128] In the case of the Lahore Ring Road, the World Bank storyline, 'infrastructure is the wheels of economic activity' was quoted several times in support of the Ring Road.[129] Therefore, the World Bank storyline provided strength to 'the Ring Road would bring economic growth' storyline.

The dominant storyline that 'the Ring Road would bring economic growth' is supported by two other sub-storylines. The first is similar to the engineers' storyline, 'elimination of congestion'. This storyline is supported by the argument that free flow of traffic is necessary to save time and energy and spur economic growth. For example, the benefit of constructing one interchange on the Ring Road was estimated to be Rs68 million in 2005 and Rs350 million in 2025 by taking into account fuel and time saving, decreased vehicle operating cost and accelerating vehicle speed.[130] In recent years the story turned the focus from urban traffic to intercity freight and passenger traffic by proposing a bypass (one section of the Ring Road) that would save travel time. Many experts believe that the 'expressway is the solution to alleviate transport bottlenecks in Lahore as public transport, especially the railway, is incapable of achieving this'.[131] However, in holding these views experts do not take into consideration the continuity of policies favourable for road construction over the last many decades. The congestion sub-storyline was also supported by the World Bank while emphasizing the importance of travel time and vehicle speed for economic development.[132]

The second sub-storyline is 'employment opportunities'. For example, 'The Lahore Ring Road employs thousands of people...The Lahore Ring Road will generate more employment opportunities for the people of Lahore...will employ unskilled labour of local communities to uplift their economic standard' (collected from several newspapers). The Ring Road storylines affirm the storylines produced at national and international institutions, as for example the Sixth Five-Year Plan emphasized reducing poverty directly by road development.[133] Similarly, the World Bank stated that, 'the construction and maintenance of some infrastructure, especially roads, can contribute to poverty reduction by providing direct employment'.[134]

The Environmental and Social Tales

The environmental storylines get attention in recent documents of the Lahore Ring Road. The Ring Road Feasibility Study[135] and the PC1[136] of one of the interchanges addressed the impact of the Ring Road on the local environment. The Study perceived environment only as a 'landscaping' issue and as such sensitive landscaping proposals were made to minimize the adverse affect of the Ring Road on the local environment.[137]

The Pakistan Environmental Protection Act, 1997 makes it mandatory to conduct detailed EIA studies for road projects. A similar requirement was made compulsory to obtain funds from the provincial P&D department and the federal Planning Commission for all road projects. However, these authorities just require Part D—the shortened version of the EIA—to be submitted along with the PC1. However, experts believed that, '...a project like the Lahore Ring Road which was conceived before enactment of the Pakistan Environmental Protection Act, 1997, does not need an EIA or an IEE',[138] '...currently no IEE or EIA is functional in transport projects in Lahore'.[139] However, it was found that an EIA of one Ring Road interchange was prepared as part of its PC1. The EIA report justifies the road project on environmental grounds and claims that noise pollution of the area 'will be minimized due to the smooth traffic flow and little traffic jams' after construction of the interchange on the Ring Road.[140] In terms of air pollution, the document explained that the construction of the proposed interchange would minimize traffic congestion and result in reducing the gaseous emissions per engine due to improvement in the traffic speed, resulting in improved air quality in the area.[141]

Both documents believe that landscaping measures, signage placement, road marking and provision of street furniture would improve the 'aesthetic value' of the road and the interchange which in turn would enhance the overall aesthetic value of the area.[142] This was confirmed from the government campaign in newspapers which said that 'a modern road would be a great addition to Lahore's beauty'.[143] However, stress was placed on selecting only local plants and trees. In short, environmental concerns are said to be addressed by publicizing 'landscaping measures', 'increasing aesthetic value', 'reduction of traffic congestion' and 'increasing speed' sub-storylines for reducing air and noise pollution. The studies have totally ignored the ecological cost of the road and road traffic, and the resettlement cost of a large number of people dislocated as a result of building the Ring Road.

The NHA also broke its silence on sustainable development by stating that:

> Roads and highways do not merely represent travelling modes but indicate the economic foundations, the social interaction of futuristic planning of national integration of a country. In short, the proper planning of a road network represents a paradigm in the sustainable development of the country'.[144]

In comparison to environmental issues which still managed to attract some attention, social issues have not (even narrowly) been addressed by any document prepared for the Ring Road project. This was confirmed by one expert, who stated that, '...no social and cultural impact of the Ring Road project has been studied'.[145] Another said that, '...many road projects need resettlement of existing inhabitants but resettlement is not considered a part of EIA'.[146] Not surprisingly then there is strong opposition from the local community to the proposed route of the Ring Road who have raised the slogan of 'No Ring Road'.[147] The community estimated that the 'Ring Road is removing (dislocating) five lakh (0.5 million) residents from their houses, shops, markets, plazas, mosques, graveyards, *madressahs* (seminaries), factories etc. and this would lead to economic disaster besides being a human rights violation'.[148] However, these social issues failed to win support in the Lahore High Court which decided that it 'does not have any right to interfere in the government's administrative and policy matters'.[149] Thus, the resettlement and human rights' violation storylines emerging on account of the Ring Road fail to draw attention.

The recent policy objectives for the Ring Road combined all the mentioned storylines by enunciating that the Ring Road satisfies the 'needs of major urban, industrial and trade centres, safety, stability and sustainability of transport systems and infrastructure with improved level of performance and service delivery'.[150] The planning, engineering, economics, environmental and social storylines are helpful in understanding how the Ring Road project was perceived and justified over time. One may conclude from this that every tale justified road building especially in the context of the Ring Road in Lahore. Notwithstanding the truth or otherwise of these storylines, and whether these were supported by evidence or not, the study can argue that the Lahore Ring Road became discursively path dependent on the mental models of both international and local decision-makers. Exploring storylines helped to explain the perceptions and beliefs of decision-makers, and even though these beliefs are constructed on professional considerations, they are endorsed by individual preferences and become strongly embedded in institutions and their funding programmes. The study of the discursive nature of the Ring Road project has shown that it was not simple to bring change to the proposed project with revised policies, and has underlined the need to demolish ingrained beliefs and perceptions that have been identified in the history of the project.

NOTES

1. There was no documentary evidence to show any progress on the Lahore Ring Road project during the 1970s.
2. Govt. of Punjab 1973: 76.
3. Ibid., 77.
4. LDA and World Bank 1980a: 111–115.
5. TEPA 1992: 5.
6. Ibid., 6.
7. The prescribed application required to be submitted to the Planning Commission or the Punjab planning and development department to obtain funds from the ADP or the five-year plans.
8. TEPA 1992: 5.
9. See Figure 7.3.4 for location of the proposed roads and Table 7.3.3 for their functions and characteristics. TEPA and JICA 1992: 7–13, 1–14.
10. Ibid., 13–7.
11. TEPA 1992: 31.
12. Ibid., 18–19.
13. Ibid.
14. TEPA 1992: 8.
15. The point from where the intercity motorway begins.
16. TEPA 1992: 6, 13–16.
17. LDA, 2004b: 21–7.
18. http://www.nespak.com.pk/news/newsdetail.asp?id=365
19. http://www.nespak.com.pk/news/newsdetail.asp?id=497
20. 'Pakistan Development Forum' is an annual gathering of donor countries and international development institutions organized by the Government of Pakistan to seek finance for development projects.
21. LUDTS, Final Report/Volume 1-A, Urban Planning, Preface.
22. LDA and World Bank 1980a: Preface.
23. TEPA and JICA 1992: Preface.
24. TEPA 1992: 8.
25. LDA and World Bank 1980a: Preface.

26. LDA and World Bank 1980c: 117.
27. Ibid., 162–164, 169.
28. TEPA 1992: 16.
29. LDA 2004b: 21–7.
30. Govt. of Punjab C&W 2004a.
31. http://www.pmupunjab.gov.pk/index.php?option=com_content&task=view&id=17&Itemid=47
32. http://www.punjab.gov.pk/p-d/h_roads.htm
33. ADP No. 287, PII.
34. Govt. of Punjab 2006.
35. Data provided by NESPAK consultant in 2004.
36. TEPA 1997.
37. Govt. of Punjab 1973: 118–117.
38. LDA and World Bank 1980a: 130.
39. LDA and World Bank 1980b: 313.
40. Information is based on TEPA and JICA 1992: Figure 13.2.1, pp.13–6, 13–7.
41. TEPA and JICA 1992: 10-11.
42. Ibid., 13–7.
43. Ibid., 10–4.
44. TEPA 1992: 16.
45. Ibid., 11–13.
46. *The Nation*, 4 December 2004.
47. *Daily Times*, 25 November 2004.
48. *The Nation*, 19 December 2004.
49. *The Nation*, 4 December 2004.
50. One *marla* is equal to 225 square ft.
51. Nadeem and Hameed 2006.
52. Ibid.
53. Hasan 2005.
54. Owens 1995.
55. Vigar 2002: 2.
56. Dimitriou 1992.
57. Howard 1898.
58. TEPA 1992: 6.
59. Govt. of Punjab 1973: 76.
60. Ibid., 76.
61. LDA and World Bank 1980a: 115.
62. Ibid., 118.
63. TEPA 1992: 12.
64. Ibid., 18.
65. LDA and World Bank 1980c: 31; TEPA 1992: 6.
66. LDA and World Bank 1980c: 119.
67. Govt. of Pakistan, Planning Commission 1960: 277.
68. LDA and World Bank 1980c: ii.
69. Ibid., vi.
70. Ibid., 30.
71. Ibid., 11.
72. Ibid., ii.
73. Ibid., 115.
74. Ibid., xviii.

75. Ibid., ii.
76. Ibid., 115.
77. Ibid., 11.
78. Ibid., 115.
79. Ibid., xvii.
80. Govt. of Punjab 1973: 75.
81. LDA and World Bank 1980c: xviii and 117.
82. TEPA 1992: 10.
83. Ibid., 10.
84. TEPA and JICA 1992: 7–1.
85. Ibid., Figure 7.3.1, 7–8
86. TEPA 1992: 20.
87. Ibid., 23.
88. Ibid., 26.
89. Ibid., 18–19.
90. Each carriageway was proposed 11.0 metres wide highway.
91. Govt. of Punjab, C&W 2004a.
92. Ibid., 9.
93. Dimitriou 1992: 191.
94. Cohen 1996: 31.
95. Dimitriou 1992.
96. Low 2005.
97. TEPA 1992: 6.
98. Govt. of Punjab 1973: 75.
99. Chief Engineer, TEPA, City District Govt. Lahore.
100. TEPA 1992: 18.
101. Low et al. 2003: 108.
102. Steering Group 1963.
103. MMBW 1981.
104. LDA, 1997.
105. http://bbc.co.uk/urdu, 16 December 2005.
106. Former DG, LDA, City District Govt. Lahore.
107. ADP 2002-03/Sr. No. 200, ADP 2003-04/Sr. No. 172.
108. ADP 2004-05/Sr. No. 287. information provided by Deputy Director Planning, Punjab Highway, Communication & Works Dept., Govt. of Punjab.
109. Chief, Roads and Bridges, P&D, Govt. of Punjab.
110. LDA and World Bank 1980c: ii.
111. Low et al. 2003.
112. *Dawn*, 18 September 2004, Chief Minister Punjab.
113. Former DG, LDA, City District Govt. Lahore
114. Ibid.
115. Mazza and Rydin 1997
116. TEPA 1992: 25.
117. Ibid., 37.
118. Chief, Urban Development, P&D, Govt. of Punjab.
119. Govt. of Punjab 2006: 711.
120. Low 2005.
121. Director, Ministry of Environment.
122. Chief, Urban Development, P&D, Govt. of Punjab.
123. Govt. of Punjab 2004: 48.

124. *Dawn*, 18 September 2004.
125. Collected from official statements published in several newspapers.
126. *The Nation*, 19 December 2004.
127. Low et al. 2003: 105.
128. Vigar 2002: 109.
129. World Bank 1994: 14–15.
130. Govt. of Punjab, C&W 2004a: Table 21, 58,
131. Director, Ministry of Environment; Asst. Chief, Urban Development, P&D, Govt. of Punjab; Chief Engineer, TEPA, City District Govt. Lahore,
132. World Bank 1994: 21–2,
133. Govt. of Pakistan, Planning Commission 1983: 10–12.
134. World Bank 1994: 20.
135. Ring Road Feasibility Study 1992.
136. Govt. of Punjab, C&W 2004a.
137. TEPA 1992: 31.
138. Director, Pakistan Environmental Protection Agency.
139. Chief, Urban Development, P&D, Govt. of Punjab.
140. Govt. of Punjab, C&W 2004a: 22.
141. Ibid., 22–3.
142. TEPA 1992: 32; Govt. of Punjab, C&W 2004a: 25.
143. *Daily Times*, 1 December 2003.
144. *Dawn*, 21 September 2004, NHA Advertisement.
145. Chief, Roads and Bridges, P&D, Govt. of Punjab.
146. Director, Pakistan Environmental Protection Agency.
147. *The Nation*, 4 December 2004.
148. Ibid.
149. *Daily Times*, 25 November 2004.
150. Govt. of Punjab 2006: 711.

8

Institutional Barriers to Sustainable Urban Transport

This chapter sums up and discusses the evidence presented in the preceding chapters, assesses the validity of the path-dependent approach to policy analysis and explores the implications of the findings for future urban transport policy making. This then would allow me to address the main theme of the book, which is 'to explore the institutional barriers that are creating a hindrance to the development, application and implementation of sustainable urban transport policies in Pakistan.'

Path dependence was postulated as a reason for institutional barriers. The concept of path dependence was divided into three fields:

1. The institutions field: to explore how institutional relationships, power and finance influence local and international institutions.
2. The techniques field: to explore which method was employed for analysing transport problems and generating transport solutions.
3. The discourse field: to explore transport problems as defined in the minds of local and international decision-makers.

The chapter is in two parts. The first part discusses the relationships amongst local and international institutions for transport planning and examines their role to explain the stability of transport policy over time. This discussion will be supported by examining the function of certain techniques and the relative influence of various storylines in bringing stability to the transport policy. The second part of the chapter assesses the validity of the conceptual framework in order to explain whether the path-dependent approach offers a useful way of understanding and interpreting a transport policy analysis.

The analysis of transport planning and policy in Pakistan has provided a fascinating example of the way in which policy paradigms developed historically, were inverted or changed, and then maintained over time.

Despite the unquestionable benefits that transport planning and policy brought to Pakistan, there is an inevitable social and environmental cost involved. Some concerns on these issues were addressed in recent transport policies but their full range was not acknowledged. It has been argued before that these shortcomings of transport policy can be usefully understood in an institutional perspective. The primary purpose of this chapter is to discuss these concerns—largely overlooked in the development of transport policy—in an institutional context. The study needs to determine if the question of sustainable development raised by the World Commission on the Environment and Development (1987) was indeed a central concern of policy-makers when formulating plans and policies for provision of transport infrastructure.

The debate over sustainable development and how sustainability was incorporated in transport decisions obviously stands at the heart of this discussion. Previously, environmental concerns in transport have been mainly restricted to local factors such as noise, air and road crashes, but recent concerns over global warming, acid rain and oil depletion have given importance to the regional, national and global implications of environmental degradation due to transport. The aim of this reminder of the sustainable development debate is to mark a point of departure which sets a context for various arguments and perspectives in the previous chapters. The starting point of the book was based on the argument that the transport policy system needs to respond to sustainable development, and provide opportunities for institutional change and development in which such a response becomes possible. It is hoped, however, that the discussion may flag some of the key issues and subsequent analysis which might be beneficial.

The analysis in previous chapters showed that the process of stabilization of Pakistan's transport policy can be usefully understood as discursive in nature. Specific storylines were adopted and diffused so as to secure the particular transport policy paradigm. In the mid-1990s, evidence that such a discourse was being structured around the environmental tale was reflected in the Lahore Ring Road case. Despite such evidence, transport policy continued to emphasize road building in the country as was affirmed by the arguments shown for promoting the Ring Road in Lahore. In fact, environmental storylines were used to justify the Ring Road, claiming that the project would have the effect of reducing air and noise pollution in the city by alleviating traffic congestion. In short, Lahore's wider concern for ecological issues did not translate into the transport policy sections of plans, strategies and projects.

The initiative of the Light Rail Transit (LRT) project in Lahore in the early 1990s was not a response to environmental concerns, but a capital intensive public transport project proposed by JICA. This initiative was altered into the Lahore Rapid Mass Transit System (LRMTS) project in the mid-2000s. While the LRT and LRMTS showed a basic understanding of the environmental benefits by promoting public transport, they failed to match the economic and social needs of the city. Perhaps it showed a weak commitment or misinterpretation of sustainable development by JICA. Some attempt was being made to introduce a franchised bus-based public transport system in Lahore in parallel to over-engineered road projects, but there was no policy to reduce traffic levels in Lahore. In fact, the federal government's car leasing policy resulted in a sharp increase of car ownership in the city. Despite some opposition from people displaced by the Ring Road in Lahore, the public at large are prepared to accept (support or be silent) individual road building projects. The increasing congestion on the road network of Lahore compels the public to accept the government's argument for building more roads. In contrast with the road project, transport institutions as well as the public have not paid sufficient attention to the development of local transport policy and planning in Lahore. In reality, the public have no say in the development of transport policy and planning. Thus, while the environment was recognized as an issue in national and local transport policy, it was concurrently used in arguments to sustain the stability of existing transport policies. Only a very restricted view of 'environment' such as landscaping alongside roads was adopted in mitigating the environmental impact of the roads.

The fundamental message of Mitchell and Rapkin is that a transport infrastructure proposal should not only be evaluated on transport criteria, but also on land-use, social and environmental grounds.[1] It seems that these criteria were overlooked in the case of the Lahore Ring Road where the detailed proposal was mainly justified on the basis of transport objectives such as alleviating congestion, increasing traffic speed and accommodating future traffic. However, the implications of the project for land use and development were not sufficiently evaluated. Similarly, the social and environmental impact of the project did not receive full attention. The investment in the Lahore Ring Road was justified to accommodate forecasted future traffic. However, the work of Flyvbjerg et al., Pickrell and Richmond has shown that there is great room for error in urban transport planning which can result in overestimating future traffic.[2] The argument of Flyvbjerg et al., Pickrell and Richmond can be applied to the Lahore Ring Road as well. Chapter 7 found that the Ring Road was initiated in the early 1960s in Lahore as an 8-lane, grade-separated, limited access highway. Rapid growth of urban development and traffic in Lahore was anticipated to justify the project. However, time has shown that this assumption and forecast were overestimated at that time. Until the late 1990s, 80 per cent of Lahore's population lived within a 7 km radius accomplishing 60 per cent of their trips on non-motorized modes (see Chapter 6). Moreover, the report of the British Standing Advisory Committee on Trunk Road Assessment (SACTRA) has shown that a new road is likely to generate more traffic instead of accommodating an expected increase of traffic based on demand.[3] Hall and Banister found the shortcoming of traditional transport analysis to be especially an overestimation of: future traffic, time savings and even the cost-benefit ratio of the project.[4] They stressed the need to conduct fresh research for developing new methods of analysis for transport investment decisions. However, the lack of research, education and the public learning process in general, and in relation to transport institutions in particular, were identified as some of the factors that resulted in Pakistan adopting traditional transport planning methods.

Although local institutions are nominally responsible to research and find solutions suited to local transport needs, they are unable to exert real influence over policy in the presence of international development institutions that are the main source of funding in Pakistan. The World Bank, ADB and JICA have helped local transport organizations but these organizations have become critically dependent on international institutions to generate transport solutions for the country, and more specifically for Lahore. When transport solutions were produced by international development institutions and their associated foreign consultants, it became the task of local institutions to arrange finance for their construction. Sometimes funding was easily received from international institutions, but when the costs were high it took decades to obtain the necessary resources. The Lahore Ring Road case has highlighted this fact. Furthermore, it was noted that once a project was proposed by international development institutions, it was difficult to overlook it in any future studies. Historically, these projects were publicly promoted on economic grounds because these are of vital importance for a developing country like Pakistan.

In further examining the argument in the above introductory paragraphs, the path dependent categories of institutions, techniques and discourse are revisited in order to explain the continuity of the specific transport policy in Pakistan. These organizing ideas provide the framework for the discussion that follows.

EXPLAINING POLICY CONTINUITY IN TRANSPORT POLICY MAKING IN PAKISTAN

The Importance of Institutional Relations in Explaining Policy Formulation and Stability

One of the aims of this book is to understand how institutional relationships, power and finance are significant in the development of transport policies in Pakistan. To achieve this aim, the path dependent components of critical juncture and self-reinforcement are introduced in order to explain policy change and policy stability in the context of international-local relationships.

a) Policy Change during Critical Juncture

Chapter 3 showed how a new paradigm of road development emerged after the involvement of international development institutions during the period of critical juncture (1955–65) despite the existence of an inherited railway network as an alternative for intercity passenger and freight transport. As discussed in Chapter 3, Thorner believed that the new railway policy was enough to enable rapid economic progress in Pakistan, as an alternative to adopting the new road development paradigm.[5] However, this did not happen and road development assumed a principal position in Pakistan's transport planning.

Chapters 3 to 5 have shown that a large number of interrelated factors inevitably underpin policy change during the critical juncture, and no single factor is likely to explain all situations. All these factors are important in explaining the outcome of national and urban transport policy today.

One of the main factors which became the turning point for transport development in Pakistan was the involvement of international development institutions and their associated foreign consultants in the First and Second Five-Year Plans (1955–65). This is a fair statement although as Chapters 3 and 7 have shown it took many years for the road development paradigm to emerge fully in Pakistan's national and urban transport policies. This change came about due to the large amount of financial and technical assistance available from international development institutions. This large scale assistance was not enough to by itself bring about a shift in the transport policy; hence it was connected with other factors to provide the necessary justification to affect policy change. In examining such factors, these issues can be divided into the institutional and the policy streams.

The main institutional issue came first because it provided the opportunity for international development institutions and their associated consultants to intervene in transport development in Pakistan. The issue was the lack of professional capacity amongst local institutions to indigenously formulate and implement transport planning and policy, although this is understandable in the context that Pakistan was then a newly independent country. However, Pakistan did not inherit a blank slate at the time of independence—the British did indeed leave behind a set of transport laws, policies and infrastructure. Local institutions sought international help to develop policies that would suit the requirement of a newly independent country. However, analysis in the previous chapters showed that instead of using the potential of existing structures international institutions laid the foundation for replacing all inherited policies and infrastructure. There is no need to invoke conspiracy theories to explain this intervention but more questionable is the failure of international development institutions to build local transport planning capacities. But even if they had acted to do so, chances are that the institutional space would have been filled with the very knowledge, techniques and organizations suited to the road-building paradigm that was dominant worldwide.

Policy issues arose from the aspiration in the late 1950s and early 1960s for rapid economic growth. Lack of adequate transport infrastructure was considered the main hurdle to rapid economic growth; hence roads were promoted as the most suitable means for overcoming this hurdle. This widespread thinking about transport policies during the critical juncture resulted in construction of roads regardless of the fact if they were justified by the growth of traffic. However, this thinking was limited to the question of developing intercity freight transport. As a result, a major portion of finance was allocated for developing roads rather than the existing railway; but perhaps this was also helped by the availability of generous amounts of international funds for developing roads and not the railway (see Figures 3.5 and 5.1). The absence of the appropriate financial capacity of local institutions and ready availability of foreign assistance for road development provided international development institutions the opportunity to get involved in transport development of Pakistan, reorienting the direction of future transport policy.

Chapters 6 and 7 have shown that this involvement and its supporting paradigm was not limited to national transport but also penetrated urban transport development. In contrast with intercity transport the policies for urban transport development were more complex in nature. Chapter 3 demonstrated that the First and Second Five-Year Plans (1955–65) focused to some degree on establishing a road-based public transport network in major cities of Pakistan. In fact, these Plans had taken initiatives to establish the [West] Pakistan Transport Board and the Karachi Road Transport Corporation, the latter responsible for running bus-based urban public transport in Karachi. Furthermore, the Second Five-Year Plan (1960–5) envisaged the construction of the Circular Railway as a rail-based public transport project in Karachi.

The development of rail- and road-based public transport in Karachi highlighted the emphasis of the federal and international institutions on national development plans, although such projects were also pursued alongside major road construction initiatives in many cities of Pakistan. For example, the Master Plan for Greater Lahore (1965–80), which was also prepared with the help of international institutions, envisaged the construction of a circular railway but at the same time proposed the construction of the 'Ring Road' even in the absence of concrete traffic data to justify the road proposal. The discussion in the preceding chapters has shown that the circular railway proposal was just a 'lip service' project while the Ring Road proposal was implemented over time. In fact, the Ring Road in Lahore was over-designed not only in the 1960s but also in the 2000s. Some sections of these road projects are related to the National Highway (N-5) which connects various cities of Pakistan and for which responsibility lies directly with the NHA at the federal government level. Others were carried out by the C&W and TEPA at the Punjab and LCDG level respectively. These latter projects were financed by the federal government through grants or credit from the international development institutions. In contrast, some sections of the Circular Railway in Karachi were built, operated and then shut down. And no serious effort was made to implement the proposal of the circular railway in Lahore. However, the JICA took the initiative in the early 1990s to propose an LRT project in Lahore which will be discussed in the following sub-section.

b) Policy Stability Over Time

The above discussion has shown that the influence of international development institutions was a key factor in transferring the road-based transport paradigm during the late 1950s and early 1960s. However, this paradigm evolved and developed through the decades up until the 2000s despite the emergence of an environmental agenda at international, national and local levels. In contrast, the money allocated for railway development as an inter-urban transport, and public and non-motorized transport at the urban level decreased.

During this period international development institutions were in a key position to influence policy as they provided over 80 per cent of the country's development budget. Despite this overall financial support, by the early 1980s most major urban road construction projects (e.g., the Lahore Ring Road) were immobilized due to the unavailability of finance from international development institutions. However, a decade long PUDP (1988–97) financed by the World Bank mobilized local institutions not only to improve the geometry of existing roads in Lahore but also to advance the long-standing proposal of the Ring Road. In the past, the transport sector has typically accounted for about 15 per cent of the World Bank's portfolio and generated an estimated rate of return[6] on completed projects of 22 per cent, compared with an average of 15 per cent for all Bank projects.[7] The critical question here for the Bank is whether the same rates of return might have been achievable if the Bank had supported sustainable transport.

In relation to inter-urban movement, road building was obviously the dominant paradigm in the 1990s. This led directly to a major road building programme including motorway and road rehabilitation such as dualization and improvement of N-5 which expanded the trunk road network. Chapter 5 charted the stability of road building policy from 1990 to 2005 and offered some explanation of its strength. It clearly showed that the existing national network of railways along with urban active and collective modes of transport were unable to disrupt the road building policy.

On the other hand the federal government has never denied the importance of public transport in Pakistani cities. The government shifted the responsibility for public transport to provincial governments after the critical juncture. However, the federal government shared the responsibility for road building with the provinces where the cost of the project was beyond the capacity of the provincial and local governments. As the capacities of transport institutions at the provincial and local levels were clearly deficient, the federal government either provided finance for building roads or built roads on their own account. After the creation of the NHA in 1991, the latter option was adopted quite frequently. In contrast, the federal government has neither provided finance nor operated by itself a decent public transport in urban areas nor improved the capacity of the provincial transport authorities. However, the federal government at one point helped provincial governments transfer the operational responsibility for public transport to the private sector. In the absence of human resources, coordination, research and financial capacity of public transport institutions in the Punjab, public transport is now fully controlled by the private sector. The incomplete routes, high fares, fewer buses, gender discrimination and even the absence of buses in some places are common in the urban areas of Pakistan. The provincial and local transport institutions in Lahore are powerless while the federal government works as an agent of international institutions in the entire transport planning process. Thus, despite this broad undermining of responsibility, the Punjab was regarded as the innovator in transport planning in the early 1980s and late 1990s. In the early 1980s, the provincial government had worked with Volvo International to introduce a 'Model Urban Transport System' in the large cities of the province including Lahore.

The Punjab Road Transport Corporation (PRTC) was created to manage the Model Urban Transport System, along with intercity public transport.[8] The main functions of the PRTC were to provide: an efficient, adequate, economical, properly coordinated public transport service; manage bus-stands and related amenities; purchase, manufacture, maintain and repair buses and allied services in the Punjab province.[9] To fulfil these functions, 650 new buses were introduced in urban areas supported by central repair and body building workshops, central stores, offices and a transport training institute located at Lahore.[10] PRTC had always been short of finance due to lack of support from provincial, federal and international institutions for a publicly owned and operated bus system. Eventually, this bus-based public transport system was closed down in the mid-1990s and the PRTC abolished. In the late 1990s, the transport department introduced the franchised public transport planning concept in major cities of the province. Lahore provides an example of a city where a franchised public transport practice emerged and was successful. In contrast, franchised public transport failed and was wound up in Rawalpindi and Islamabad. In fact, no one knows how long this practice will last in Lahore because the institutional capacity of the transport department to support it was not strengthened. Moreover, the provincial government failed to attract the attention and assistance of the federal government and international institutions for this project because the latter were more interested in building roads in Lahore, arguing that new roads help to increase the speed of bus-based public transport.

In terms of rail-based public transport, the LRT proposal appeared in 1992 in the CSTS, studied and funded by the JICA.[11] The project was very capital intensive and relied completely on technical and financial assistance from the JICA. The study justified the viability of LRT operation by proposing relatively higher fares. Badami describes the results of surveys in Indian cities which show that the financial viability of urban rail systems depend critically on large populations with high per capita income to provide the revenue base to sustain them, alongside high utilization and fares with low staffing and wage levels.[12] However, people in Pakistani cities are predominantly poor; the per capita annual income in Lahore is only US$800–1000[13] which is not enough to support LRT fares. Lahore's lower per capita income makes it unlikely that the city could support the LRT without massive subsidies which is something that the country cannot afford.

The LRT will achieve savings in traffic accidents and some reduction in air pollution, but the proposed grade-separated structure will itself have a major impact on the urban scene of the historic and cultural buildings of Lahore. Barter et al. argue strongly in favour of urban rail because of its potential to transport large numbers of people quickly, reliably, comfortably and safely, and its ability to wean people away from personal motor vehicles.[14] This is true for Hong Kong and Singapore which, despite their wealth, have low automobile dependence and high public transport usage. While both these cities have the demonstrated ability to control land use, to plan and implement intense mixed development integrated with rail-based public transport, and to apply strong economic and traffic restraint measures to curb personal motor vehicle ownership and use, it is unclear how successful Pakistani cities will be in applying these controls. Therefore, the Hong Kong-Singapore model cannot be applied to Lahore because of the higher income level and strong institutional capacity of the former to run the system. However, Pakistan needs to learn the lesson of Singapore's successful public transport system which is not properly described without reference to its battles in the early 1970s with the World Bank and some American consultants who advised Singapore against developing an indigenous urban transport system that was suited to its local characteristics.[15]

In fact the LRT is simply a plan advanced by international development institutions—this time it was the JICA—to introduce an alternative form of public transport. Newman and Kenworthy argue that a public transport system should be implemented in conjunction with restraints on car ownership and a commitment to maintaining traditional forms of non-motorized mobility.[16] They posit that these methods are effective in dense, mixed land-use settings with short travel distances for many trips. Therefore, strong economic controls to curb personal motor vehicle ownership and use by means of imposing high taxes, parking costs and traffic restraints would be required. The success of the urban rail system in Hong Kong, Singapore and Tokyo is in fact largely due to these features. However, the Lahore Transportation Master Plan prepared by the JICA only proposes a public transport system without formulating proper policies for other sectors mentioned by Newman and Kenworthy. In the absence of all these supporting and complementary policies in Lahore, it is doubtful if the LRT system can succeed. In contrast to the JICA, the World Bank emphasized the development of a Bus Rapid Transit (BRT) system in Pakistani cities.[17] However, they focused on the need to have intensive management of existing road space along with an improved capacity of key junctions before introducing a BRT system in Pakistan.

Sustainable transportation options rely heavily on the promotion of affordable public transport and non-motorized modes.[18] For low income people commuting to work, walking, bicycling or affordable public transport are not a matter of choice but a necessity for survival. Therefore, it is critically important that the urban transport network be designed from the perspective of the pedestrians, bicyclists and low-cost public transport vehicles that provide convenience, and are safe, comfortable and affordable. However, the policies that are actually promoted do not recognize the importance of non-motorized modes, even actively discriminating against them by ignoring their importance in urban transport.[19] In addition, decision-makers pursue transport policies that work against walking and cycling. This is despite the fact that these modes can contribute significantly in improving environmental sustainability. Instead of emphasizing pedestrian planning as in many Western cities, particularly in Europe, Lahore urban transport policies and programmes have not accommodated pedestrian and other non-motorized transport in a future urban transport scenario. In the presence of high density, mixed land use and the tradition of using non-motorized mode which generates a large number of trips in Lahore, the prospects for effective pedestrian planning are more favourable than for any other project, if only the facilities existed for them. Even bicycle travel can be promoted by providing adequate infrastructure. Although funding levels and financial capacity of local urban transport institutions are deficient, yet the major portion of the available money has been devoted to large-scale roads projects.

What Lahore needs is small-scale urban transport improvement and management, a reasonable share of finance devoted to non-motorized transport and public transport services, and an infrastructure integrated with an improved intercity and suburban railway system. Institutions in Pakistan need to develop transport policies that take into consideration Lahore's unique needs and priorities as well as the country's capabilities and financial constraints to achieve low-cost, safe and robust policy outcomes. However, locally designed development programmes, especially in urban transport, raise questions of local institutional capacity. Dimitriou explained that institutions can deliver sustainable development if they have the,

> ...capability to be involved in the planning, management, operation and maintenance of development efforts to sustain themselves both politically and functionally. [Moreover they] should be able to carry out their current and future tasks efficiently and effectively, and [be] capable of generating and renewing their manpower, financial and information resources on a continuous basis.[20]

Dimitriou's concept also 'conveys the presence of qualities of good coordination, transparent decision-making, clear lines of accountability and responsibility, community participation, equal opportunities and access to competence.'[21] By applying all these characteristics in local transport institutions in Pakistan, the study mainly found inappropriate planning, management and operational capabilities; weak human and financial resources; ineffective political and functional competence; lack of information and research aptitude; absence of coordination and transparency and reluctance of community to participate (see discussion in Chapters 5 and 6).

The widespread overlapping of responsibilities in urban transport institutions in Lahore suggest the need to overhaul institutions to provide clear job descriptions, unambiguous coordination procedures and a staff with appropriate skills. Moreover, this study recognizes that political and economic stability which is vital for building transport institutions and infrastructure is lacking in Pakistan. Even though institutional issues were of a very serious nature, international development institutions mainly focused on the development of 'hardware' (technical and infrastructure investment) and paid no attention to the 'software' (human investment). It would seem that the international institutions did very little to build transport planning capacities of local institutions as was promised in the First and Second Five-Year Plans (1955–65). There is no need to invoke conspiracy theories to explain this incidence. However, this lack of attention helped to stabilize the road development paradigm over time. The main concern of the study is that local transport institutions played so small a role in the transport planning of Pakistan which was mainly driven by international development institutions as part of a knowledge-cum-technology transfer phenomenon.

c) International-Federal-Local Relationships in Explaining Policy Formulation and Stability

In their study of the central-local relationship in the UK, Wilson et al. found that the local government usually works as an agent of the central government.[22] These authors found that through control of financial resources, the central government is in a position to exert considerable authority over localities. Moreover, the central government sets the regulatory and policy context for localities in the UK. The situation is similar in Pakistan but in international-federal-local relations. Because of their long standing control of financial resources for development in Pakistan, international institutions are now in a driving position to formulate the country's transport policies and programmes. The regular updating of the National Transport Plan by the JICA,[23] the recent ADB initiative to formulate the national transport policy and a similar effort by the World Bank (in the form of TSDI) are some examples that were discussed earlier. Moreover, experts from resident missions of these agencies remain actively involved in local efforts to formulate national transport policies which provide guidelines for provincial and local institutions to develop their transport plans. Overall, transport institutions at all levels of government aim to develop close relations with international funding institutions. A similar effort governs relationships amongst the country's national, provincial and local institutions. The local government of a city depends for its funding on the P&D department of the provincial government which in turn seeks similar assistance from the Planning Commission of the federal government. This chain of dependency helps to implement transport policies prescribed by the international development institutions. There is of course a future role for international institutions but this should be within a clear strategic framework which examines the problem holistically with other planning and engineering institutions, and within which international institutions' advice can be best utilized.

In the urban context, the power of international development institutions has been confirmed by the analysis of the Lahore Ring Road proposal in the previous Chapter. Such analyses also validate the argument that the road development paradigm was advanced without taking into account the pre-requisite local data. The overriding importance of economic growth and engineering led the solution, together with a failure to internalize environmental and social costs, meaning that the Ring Road proposal simply ignored the sustainability criteria. In formulating this project, international development institutions and their foreign consultants played a key role while the Planning Commission was their main partner at the national level. It would seem that to get funding for projects, the Planning Commission acted according to the norms expected of it by international development institutions. To a very large extent, provincial and local institutions also acted similarly besides which the latter engaged in a continuous dialogue with international institutions and prepared studies over time for the Lahore Ring Road project. Such stability of approach to policy was generally aided by the preferences of international development institutions. More generally, the power of international development institutions in transport development was supported partly by a centralized governance system which diminished the ability of local institutions to shape transport policy. As a consequence, local institutions have tried to develop strong relationships with national institutions that in turn have developed strong relationships with international institutions.

The structure and capacity of local transport and planning institutions are also key components in understanding the power of international development institutions and the stability of transport policy. The absence of appropriate financial, human resource and research capacities in transport institutions provide continuous openings for international development institutions to become influential in transport policy making. Foreign advice can be best utilized if indigenous institutions are strong and competent because otherwise they fail to identify methods and techniques to obtain solutions best suited for local transport problems. Such was the case in the current study of the Lahore Ring Road project (see Chapters 5, 6 and 7). The ultimate result was that local institutions could envisage no alternative to recommending expansion of the road programme for which money was readily available from national and international institutions. In the local scenario, transport policies were typically determined by departments geared to engineering (e.g. NHA, C&W and TEPA) and economics (e.g. Planning Commission and P&D). There was an obvious lack of connectivity and coordination between such departments, particularly the public transport (such as Transport Department) and land-use planning (such as LDA) departments. The internal divisional capacities within P&D and TEPA also show noticeable priorities to road engineers over public transport planners. Moreover, the road building and the environmental protection departments at all levels work in complete isolation of each other as is evident by the local EPA's decision to ban the two-stroke rickshaws in Lahore without consulting the transport department or the rickshaw union. However, there was evidence that this situation was undergoing change because there was a move to cultivate links to other policies (TSDI and SMEDA transport policies) and an increasing recognition of the public at large (MTDF transport policies). However, the role of land-use planners in formulating transport policies has not yet been recognized. The creation of a transport ministry at the national level might secure a more balanced policy, as suggested by various policy documents. The separation and lock-in thinking of various transport policy institutions, both at the local and national levels, helped the continuity of the road based paradigm after the critical juncture.

The above discussion has highlighted the structural nature of the power of international development institutions and the culture of local planning and transport institutions. The human resource capacity of planning organizations (such as the Planning Commission and P&D) is closely associated with the disciplines of economics, while transport organizations (such as NHA, C&W and TEPA) rely on the abilities of civil engineers. Neither economists nor civil engineers have the knowledge or skills required to plan an effective and sustainable transport system. The role of the Planning Commission and the P&D provides authority to economists to formulate transport policies within the financial resources available from national and international institutions. The civil engineering profession and its related organizations (such as NHA, C&W and TEPA) derive prestige from large-scale road projects and the control of large capital budgets. In contrast, the land-use planner, public transport planner, environmentalist and their organizations (such as the transport department, EPD and LDA) command few financial and human resources (see discussion in Chapter 6). In addition, evidence from the Ring Road case in Lahore highlighted the fact that projects are not integrated with land-use planning hence the priorities of civil engineers are guided by the techniques and methods that are employed for justifying transport.

The dominance of the civil engineering profession in local transport institutions has resulted in the promotion of road-based transport planning, policy and research in Pakistan. All transport problems in Pakistan are therefore viewed from the perspective of a 'single solution', that of road building programmes. A further product of an engineering-led approach was the over-engineered design of the Ring Road project. Similarly, the introduction of LRT and LRMTS in Lahore would appear to be an excellent piece of engineering (if the money is received and the project completed), but it was a very expensive solution considering that a basic structure of good public transport was non-existent. Similar engineering-led approaches can be seen in other cities of Pakistan (such as the Lyari Expressway in Karachi, Zero Point and Faizabad interchanges in Islamabad and Abdullahpur interchange in Faisalabad). Such high profile transport projects are often given attention as they have the potential to attract funding from international development institutions and ultimately become a political symbol of development. These are the projects by which governments demonstrate their commitment to development; in contrast, public transport, controlling air pollution and cycle tracks get little political and media attention.

Moreover, the automobile industry (which is mainly captured by Japanese firms such as Honda, Toyota and Suzuki), local and international petroleum companies (Pakistan State Oil, Shell Pakistan and Caltex), road construction industries (Daewoo and the FWO) and professional groups (Pakistan Engineering Council, Chartered Institute of Logistics and Transport), have a role in stabilizing the road-based transport policies. However, analysis of these is beyond the scope of the study and needs separate attention.

All the above mentioned groups are well-resourced and seem to be well connected to government transport organizations historically. In contrast, land-use planning and environmental organizations are weak, under-resourced and are relatively new entrants in transport policy development. However, environmental organizations also carry forward the agenda of international environmental NGOs (see discussion in Chapter 4) that are more interested in biodiversity and conservation rather than the ecological impact of road development. The work of Platteau has already shown that there is a serious problem of 'elite capture' in the work of NGOs in developing countries.[24] Therefore, the bias of environmental NGOs further empowers existing institutional structures. There is a notable absence of grassroots citizen or community involvement in transport policy. Some groups affected by large-scale road projects (as in the case of Lahore Ring Road—Chapter 7) try and protest against the current paradigm but the historical absence and weak voices of these groups have failed to impact the continuity of road-based transport policies.

In the early 1990s, a change occurred in relations between the private sector and the government. TSDI and SMEDA transport policies were developed to promote stronger public-private relations. The government wants more involvement of the private sector in the construction and management of roads and the operation of public transport.

There is a long history of public-private relations in the operation of public transport. Historically, the provincial governments in Pakistan have owned and operated intercity and urban public transport services. However, the Second Five-Year Plan (1960–5) advocated the encouragement of the private sector in public transport along the guidelines provided by the international development institutions. Additionally, the decline of state-owned public transport services created a vacuum which was filled by private operators in accordance with the Second Five-Year Plan. Initially, the market was open to private operators in parallel with state-owned public transport but since the latter has not grown at the same rate as the population in Pakistani cities,[25] a large number of small private operators (operating Toyota 18-seater, Mazda 25-seater and Suzuki 12-seater vehicles) have been permitted to fulfil the need in a fragmented way.[26] However, state-owned bus services were always given priority over private operators in the allocation of profitable routes and in reality, private operators are at the mercy of the provincial transport department for route allocation.

Mees argued that an improvement of 'service quality' and the 'integration' of public transport were needed for making public transport successful.[27] It has been observed that an average headway for small private public transport in Lahore was only 4 minutes, while maximum and minimum headways observed were 8 and 1 minutes respectively.[28] However, the improved headway has not provided the best 'service quality' which was also managed in a fragmented manner. As a result, a chaotic mass of individually owned small vehicles operated in urban areas, competing for road space.

A hope emerged in 1998 after state-owned bus services were terminated and the public sector became a regulator for the privatized operation of public transport. This hope was based on the expectation that private operators would be involved in the public transport policy making process to facilitate efficient public transport operations in Lahore. On the contrary, the Punjab government shut down public transport services operated by individual private operators in order to introduce bus-franchising schemes which offered exclusive rights to franchise bus companies on selected routes. This policy was formulated without consulting private transport operator unions (either buses or wagons) and without considering their established services for providing public transport in Lahore. In reality, thousands of private operators lost their livelihood because their route permits were cancelled to allow exclusive operating rights for franchised bus services.

Tiwari and Pucher et al. are in favour of privatization of public transport services to generate funds to operate efficient bus services, and to reduce government subsidies.[29] Tiwari found that large scale privatization of buses in New Delhi increased the capacity of public transport.[30] Pucher et al. reported that the privately run services in Indian cities have higher productivity, lower costs, and higher revenues per bus km of service.[31] They concluded that privatization does have the potential to improve efficiently, but it must be accompanied by institutional capacity building which ensures an integrated network of public transport services. This suggestion is very difficult to implement quickly because there is a long way to overcome institutional disco-ordination, lack of human, technical and financial capacity among transport organisations in the developing world.[32] Even in Europe and Australia, which have well-established institutions, the privatization and deregulation of public transport has been a significant factor behind the decline in patronage and services.[33]

In his study of developing African countries, Kah found that privatization of public transport became extremely chaotic in the absence of properly coordinated regulatory institutional structures and mechanisms.[34] He concluded that governments have a significant role to play in defining policy, developing national level coordination and local level implementation, accessing adequate investment, and building technical and professional capacity for public transport. This does not happen in Pakistan, more so at the local government level, with the result that all policies for encouraging the private sector to operate public transport failed over time. Haider and Badami reported that the franchised bus company in Islamabad/Rawalpindi ceased its services due to lack of capacity in the public sector to resolve conflicts over exclusive operating rights, route allocation and fares.[35] These conflicts were also present in Lahore where local and foreign franchised bus companies demanded government subsidies for providing lower fare to students and disadvantaged people. In summary, public transport was operated by the private sector (private operators with fare regulation and franchised companies with fare deregulation) since the 1960s. Yet the role of public transport operators, either the wagon union or franchised bus companies, has not been recognized in the development of any policy related to public transport in general and urban transport in particular. The situation in Lahore clearly shows that public transport planning has not been successful due to inadequacies in the overall institutional structure which does not allow private-sector involvement in the policy making process.

In terms of public-private dealings in road projects, the construction of the M2 (Lahore-Islamabad Motorway) is one of the examples of their relationships. In the case of the Ring Road, local transport organizations concentrated on encouraging the private sector when the Lahore Transport Project was dropped by the World Bank. However, the result, as Banister found that, 'the private sector is reluctant to invest in transport infrastructure in most of the world, even in partnership with the public sectors' except when monopoly profits are guaranteed.[36] As the traditional form of public sector finance has been limited over time, this project awaits strong intervention by the World Bank. In short, the relationship between private and public institutions appears again to be a facilitating factor for the stability of road-based transport policy in Pakistan.

To summarize, the dominance of the engineering-based human resource in local transport organizations explains the configuration of the existing transport policy. Such local transport institutions were strengthened over time and were able to impose their thoughts on the national and local transport decision-making processes. Storylines such as 'roads for development' were vital in bringing these institutions in synch with international institutions. Working continuously together over a period of time helped develop a working relationship based on shared logic and formulas that justified their thinking. These thought processes became apparent in the transport solutions, forecasting and appraisal methods of transport institutions. Alternative views were viewed through the lens of the dominant thoughts. The environmental tales explained in the previous Chapters are good examples to support this argument, when environment was used simply to justify various road projects. In fact, the discussion on the National Conservation Strategy in Chapter 4 showed how NGOs defined protection of the environment merely in terms of mitigating air and noise pollution in urban areas. Thus, narrow interpretations of environment were incorporated into the existing transport paradigm to suit the dominant view. In this manner transport policies were developed around local and international institutions, their logic and the storylines they promoted. This process gained momentum over time and road transport became a dominant paradigm.

One interesting difference between local transport policies in Lahore in the 1960s compared with the 2000s lies in the role of local transport institutions. In the 1960s, it appears that local institutions were often unconvinced about road developments while international development institutions tended to promote them. This assertion is supported by the proposal of the Master Plan for Greater Lahore (1965–80) which included the Ring Road but also emphasized public and non-motorized transport. In contrast, local experts prepared an Integrated Master Plan (2001–2021) in the early 2000s which exclusively pushed road development schemes in Lahore. There is almost no proposal for public transport and non-motorized transport in this Master Plan (see discussion in Chapter 6). This reflects the change over time in the local institutional paradigm. So by 2005, Lahore has arrived broadly at the same position as in the 1960s: pursuing road building policies.

The main implicit assumption underlying the work of the international development institutions has been that there is just one path to 'development', that of the 'Western' nations. Yet it is clear today this path is unsustainable and in fact impossible for the whole world to adopt because it would overstretch the resources of the planet. Therefore, other paths must now be explored both by developed and developing nations. A plurality of paths must now be considered, derived not from the experience only of the developed world but from the real mobility needs of the people of different nations of the developing world.

Furthermore, the emergence of the road building agenda cannot be explained merely as top-down impositions by the federal and international institutions or a bottom-up revolution by the local government. In fact, it was not only driven by the federal and international institutions but also embedded in mental models and lock-in techniques. It also reflects the fact that the road building paradigm is now embedded deeply in local institutions; even local experts saw congestion being potentially relieved by building more roads. Of particular importance were the specific professional human resource capacities of existing institutions, as described earlier, that provided the channel for spreading these mindsets. This state of affairs requires an alternative strategy, resourced both locally and globally, that can boost development by using sustainable planning and management approaches. The above discussion also highlights the role of analytical techniques and discourse in path dependent transport policies which will be explained in the following section.

The Importance of Techniques in Transport Policy Stability

One of the aims of the study is to explore the different methods and techniques that were employed for analysing transport problems and generating transport solutions to assess whether or not transport planning is locked-in to a rigid technical field in Pakistan. Chapter 5 explored different techniques which provided stability for the road development paradigm in intercity transport in Pakistan, while Chapter 7 explored the techniques used to provide stability of transport policy in general, and the Lahore Ring Road in particular. The purpose of this section is to highlight the findings of these chapters and discuss how these techniques helped to promote the road development paradigm in Pakistan.

a) Method/Techniques for Intercity Transport Development

Economic and traffic forecasting and the application of road density formulas were the main techniques that were used for intercity road development of Pakistan. Chapter 5 explained how 'grow, guess and gain' was used in the earlier Five-Year Plans; the technique is based on the assumption that GDP growth would bring a higher volume of traffic which would ultimately require more transport infrastructure, mainly roads. Over time, new roads brought an increase in traffic and road construction became an inevitable solution. These trend-based forecasts helped economists in the Planning Commission and P&D to allocate financial resources for roads and railways. Due to a decline in rail freight and passenger services and the increased role of roads and road traffic, financial allocation favoured road development starting from the Second Five-Year Plan (1960–65). In spite of a large amount of resources spent on road development, the traffic did not increase as forecast. This deficiency created the need for another technique or formula to justify the investment on roads.

Therefore, to justify investment in roads a crude indicator of development, i.e., 'road density' (road length in kilometres per square kilometres of area) was created. It was assumed that economic growth required 0.5 kilometre of road length per square kilometre of area, irrespective of traffic volume. The 'decide and provide' formula was adopted by the West Pakistan highway department in the late 1960s, while the Planning Commission officially adopted this criterion in the

Fifth Five-Year Plan (1978–83) to make up for deficiencies in roads in the country as a whole. From then, 'decide and provide' was adopted as the main technique for justifying road investment in Pakistan.

On the basis of this technique, road development extracted most of the public funding in Pakistan. Data of different countries shown in Table 5.1 does not support the claim that higher road density brings greater economic growth. Moreover, it is important to recall that the four provinces of Pakistan—the Punjab, Sindh, NWFP and Balochistan—have large variations in area and population. For example, Balochistan comprises 43.6 per cent of the total area of Pakistan with just 5 per cent of its total population.[37] Similarly, the Punjab comprises 25.8 per cent of the country's total area but has 55.6 per cent of its population. Table 5.1 clearly shows that Pakistan has attained comparable road density with many high- and middle-income countries, but there are no indications of an increasing GDP per capita in the country. In fact, it was growth that brought high road density in these countries rather than higher road densities enhancing growth. Although roads are necessary in some areas of economic growth, other forms of transport can also serve this purpose. There are a variety of human needs connected with mobility which require a plurality of transport paths. The uniform application of such a crude formula for road development in a country that has huge demographic and geographical variations resulted in the using up of a large proportion of Pakistan's scarce resources. The formula was developed by American consultants who worked in Pakistan in 1962 and was formally adopted in the late 1970s by the Planning Commission. In Pakistan, as elsewhere around the world, scarce resources should be wisely used to maximize benefits for the largest number of people.

b) Method/Techniques for Urban Transport Development

In contrast with intercity road development, forecasting and modelling practices embedded in traditional land-use planning were the major techniques that were used for justifying road development in urban transport planning in Lahore. These were enrolled into land-use planning with the 'predict and provide' or UTPP techniques.

The initiative for the Ring Road shows how planners perceived land-use development in Lahore on the basis of a popular design and aesthetic concept of the Garden City—having suburban, low-density housing, green belts and new orbital and peripheral roads. However, in contrast to the full design and institutional concept of Ebenezer Howard's *Garden Cities*,[38] the main concern of the Master Plan for Greater Lahore (1965–

80) was provision of low-density housing served by roads with grade-separated interchanges, limited access highways and the Ring Road. In fact, the notable concept was of garden suburbs, not a garden city, aiming to decentralize housing away from the limits of the old city of Lahore where jobs were located.

According to Hall, the concept of garden suburbs was developed in the early twentieth century at Totterdown Fields in the London County Council (LCC) area, Hellerau in Germany, and at Forest Hills Garden in New York.[39] Sussman explained how the Regional Planning Association of America (RPAA) in 1923 effectively married the ideas of Howard, Kropotkin and Geddes into a central concept of decentralized communities with segregated land uses and open space accessible by private automobiles.[40] In fact, the Master Plan for Greater Lahore (1965–80) was influenced by these concepts which were already being practiced in the developed world. However, due to lack of resources in Lahore the focus was to develop 'self-build sites and services' for private housing in low-density suburbs, totally ignoring the institutional aspect of the Garden City. According to Hall, 'self-build sites and services' housing was the accepted belief in the World Bank, which looked to provide low-cost and effective alternatives to government-organized public housing schemes in the developing world.[41] Due to the involvement of international institutions and their foreign consultants in the entire planning hierarchy of Pakistan, these planning concepts were transferred to and practised in Lahore. Therefore, the Lahore Ring Road project was justified on the grounds that it would serve the low-density private housing conceived in the Master Plan in the early 1960s. In spite of allocation of large resources, the building of transport infrastructure and support of national and international institutions, low-density housing areas of Lahore contain less than 20 per cent of its total population.[42] The remaining 80 per cent of Lahore's population still lives in the inner and intermediate areas within an average 7 km radius circle, maintaining an average of medium to high density of 150–250 persons per hectare.

Besides planning concepts, the notable similarities among all planning and transport studies in Lahore were that all rely on conventional transport techniques (UTPP or the 'predict and provide') and concepts developed and practised in advanced countries. The broad structure of UTPP followed that of the Chicago Area Transportation Study in the 1960s with minor modifications and assumptions.[43] Vigar called these assumptions a 'black-boxing' necessary to promote specific models and techniques for the 'predict and provide' paradigm in the UK.[44] In the case of Lahore, the modification or assumption was emphasis on GDP growth rather than traffic forecasting that is used in many developed countries. It was believed that the expected GDP growth was the basis on which to forecast future vehicle and passenger volumes in the main corridors, and an increase in road and public transport capacities was proposed to accommodate these expected increases over the following twenty years. The four-stage aggregate model, as originally developed in the USA, has been the foundation of all Ring Road and LRT analyses. Banister found that the basic structure of the UTPP has proved robust because of which it has spread throughout the world.[45] However, he argued that the theoretical foundation of the UTPP is weak compared to its empirical basis, but more specifically it does not answer the question why do people need to travel. It also contains no concept of serving existing human needs for mobility and the capacity of people to pay for travel.

Banister argued that besides the different variables used in the UTPP, there are strong assumptions regarding quantification of variables such as the value of time and life, the value of the environment and even the length of a trip rather than component parts of the trip.[46] Naess explained that neither the time saving nor the environmental impact are easy to predict accurately and values these variables in economic terms.[47] Due to the complexity of human factors it is not easy to factorize them in this technique; therefore the utility of UTTP for decision makers is that traffic forecasts are overestimated in favour of road construction at the cost of alternative strategies.[48] Naess argued that if a road construction project shows financial gain in terms of saving time, why is it predetermined that society at large is willing to pay for this benefit.[49] He raised several questions about the utility of the UTPP and cost-benefit analysis for supporting road investment projects. He argued that attempts to investigate assumptions of continuous economic growth, consumption of unlimited natural resources, time saving, distribution of benefits and costs amongst different population groups and willingness-to-pay encountered serious problems. In the light of these arguments, the National Audit Office's investigation into road planning in the UK (1988) already acknowledged that traffic forecasting is not an exact science and there are many sources of possible inaccuracies such as human and measurement errors, statistical and modelling errors, prediction errors on specific routes, errors in GDP and fuel price changes and social errors.[50] This critique became more apparent after the SACTRA report[51] which said that adding to the road infrastructure can provide short term relief to congestion but brings additional traffic in the long run. However, Morisugi and Hayashi found that the technical aspect of the UTPP and the cost-benefit analysis were in fairly common use in the UK, France, Japan, USA and Germany in spite of significant differences in their institutional setups.[52]

The UTPP technique used in Lahore supported road development projects but lacked an overall vision of transport in the city. Dimitriou found that the UTPP and its derivatives are more supportive of private motorized transport instead of the non-motorized transport that is common in the developing world.[53] Figures for Lahore show that walking is one of the most important (comprised of 51 per cent trips) yet neglected modes of travel in urban transport planning in Lahore. Travel demand in Lahore is strongly oriented to short journeys mostly covered by foot (less than 2 km) yet the UTPP fails to give appropriate value to walking trips. Over all, UTPP failed to accommodate the large proportion of pre-existing transport modes and non-motorized transport in Lahore. Clear evidence of this is seen in the master plans which neglect or under-emphasize the 'lowerarchy' (non-motorized transport facilities) but aggressively push the more expensive highways (the Ring Road) and rail (LRT) investments. The main concern of transport planning has been to increase the quantum of travel (mainly cars) but its quality has been ignored (mainly pedestrian and non-motorized transport). The Ring Road project was based on the belief that the road would meet the demands of expected future traffic levels that would be determined through careful quantitative analysis. However, the main drawback of this technique used for justifying the Ring Road and LRT proposals was an unrealistic approach to investment in a large-scale transport infrastructure that is totally mismatched with the needs of the vast majority of the people of Lahore.

Additionally, the UTPP totally ignored the political scenario which is important in Lahore, and also underestimated socio-economic and environmental considerations. Despite the existence of a legal basis, guidelines, regulations and a hierarchy of environmental institutions, environmental considerations were not incorporated in the UTPP nor were separate EIAs conducted. There is a possibility that the government might conduct EIA studies for some sections of the Ring Road in future. However, Eade's research on one freeway (expressway) project in Melbourne shows that an EIA fails to provide information to decision-makers about the environmental impact and possible alternatives for such a large and complex transport infrastructure project.[54] Equally, the social impact of the Ring Road was ignored in the UTPP as the practice of Social Impact Assessment (SIA) is non-existent in Pakistan. Furthermore, while the whole UTPP process justified the Ring Road, it did not give attention to a coherent planning framework. Dimitriou found that urban transport planning requires a multi-disciplinary approach as compared to conventional engineering science which is especially inappropriate to developing countries.[55] However, the experience of Lahore shows that this principle has not yet been recognized by those who plan and formulate local transport policy. The UTPP was embedded in the practice of international consultants and institutions which spread to local institutions offering transport solutions for Pakistan. Building roads for economic development, and later to resolve congestion, has always been seen as the most significant issue that requires attention. This view was supported by the centralized institutional structure in Pakistan. Thus, the international institutions found allies in local organizations such as the Planning Commission and the NHA at the national, the P&D and C&W at the provincial, and the TEPA at the local level, with shared worldviews and similar dependence on financial and technical assistance for road construction. There was a two-way link in which local organizations depended on federal organizations and federal organizations depended on international institutions for financial and technical resources. Similarly, international institutions depended on federal and local institutions to implement the road development paradigm. Over time, these links have been strengthened and have stabilized road development practices in the country by employing specific formulas and techniques developed by certain professions.

In short, the Ring Road in Lahore was promoted to achieve a high level of efficiency for rapid economic growth, while land-use planning and environmental protection efforts, particularly in Lahore, were almost completely ignored in the name of providing an efficient transport system. Dimitriou argued that conventional techniques have a single aim—to attain transport system efficiency—and this supersedes all other objectives of transport.[56] This is so because UTTP techniques have become well equipped over time with tools and methods to design and plan operationally efficient transport networks while the equivalent expertise in the planning and management of more sustainable transport systems is far less advanced.

Dimitriou argued that the critical technology transfer question is not about the techniques of urban transport planning but about the strategy which cities in developing countries should not replicate from developed countries.[57] This study found a different picture in the case of Lahore. By and large, the transport sections in master plan studies in Lahore are usually no more than a compilation of investment programmes for road building and improvement, and small-scale traffic management derived from the UTPP forecasting exercise. Due to the lack of capacity in local institutions, the strategy could not get full attention and the techniques for identifying large-scale transport projects became increasingly more important. The techniques carry weight because they are quantitative and look scientific. It is possible to develop equally convincing quantitative techniques around the goal of sustainable transport. But they have not yet been developed or disseminated.

In sum, the UTPP does not have enough capability to adopt a multi-disciplinary approach and to address complex transport situations in developing countries. To address the limitations of the UTPP, a group of planners from developed countries has already concentrated on institutional issues in relation to the communicative theory, taking account of both formal and informal decision-making structures.[58] However, there is a need for such a type of research to be produced in developing countries as well.

Overall, the discussion has highlighted the fact that local transport policies and solutions in Lahore are path dependent in the technical field, and this has been a hurdle for innovative thinking required to promote sustainable transport. To address the issue of technical path dependence, transport planning techniques should undergo radical changes in order to attain the objectives of sustainable development. It is now widely recognized that the previous practice of transport planning basically originated from the highway building movement and its concerns about increasing the capacity of the system to meet the expected levels of demand. In developed countries there is now a growing consensus that roads cannot be provided to the same level to match the increase in demand, and even if it were possible to invest in expanding the infrastructure, this is not seen as desirable for financial and environmental reasons.[59] Now is a time that transport planning in Lahore should recognize the limitation of the UTPP approach and accept the fact that decisions are institutionally embedded and require a new approach to sustainable transport. In sustainable transport, the planning process has become more holistic with the linkage of transport components with housing, employment and environmental goals: there should be involvement of a wider range of institutions and affected groups, and more particularly a democratized and transparent decision-making process.

The Importance of Discourse in Transport Policy Stability

One of the aims of this study is to explore the discourse field by considering how transport problems are defined in the minds of local and international decision-makers and assess whether the discursive form of path dependence can be detected in their perceptions. Chapter 5 explored storylines mainly running around economic and environmental tales which provided stability for the road development paradigm in intercity transport in Pakistan. Similarly, Chapter 7 found town planning, engineering, economic, environmental and social storylines provided strength for road development in general and the Lahore Ring Road in particular. The purpose of this section is to discuss the findings of these chapters and to consider how these storylines helped in supporting the road development paradigm.

The last two sections suggested that institutional relationships and analytical techniques played a significant role in stabilizing transport policy after the critical juncture. However, the discussion in the previous chapters suggested that the change that occurred at the critical juncture became a more permanent shift due to discourse. The discrediting of old discourse and emergence of new discourse in the form of different storylines are key factors which made the policy stable.

After the critical juncture, transport policies were carefully delivered to promote the road development paradigm. In the beginning, attempts were made to weaken railway development by pushing the discourse for road development in intercity transport development in Pakistan. Storylines such as 'roads for development', 'economic growth needs roads', 'roads will bring economic growth' and 'road development as a vision of economic prosperity' were then able to appear as promoting possible alternatives to the existing railway. However, the discourse for road development worked with pre-existing public and non-motorized transport in urban areas. Storylines such as roads for 'missing links' and 'roads to eliminate congestion' were then able to appeal to the existing public transport organizations and users and those who used non-motorized modes. In fact, these storylines were projected to accommodate the large percentage of cycling, public transport and pedestrian trips and get wider support for the ongoing discourse.

All international and local transport and planning institutions made claims that road building would bring economic growth by allowing movement, relieving traffic congestion and ultimately saving time. The main storylines were therefore about 'employment growth' associated with road development. In the early 1990s, road development in urban and regional areas was hidden in the storylines of 'employment opportunities' and even 'direct employment opportunities for poverty reduction'. However, the NTRC transport policy in 1991 provided evidence that the assumption that road development brings economic growth was not proved in the case of the Karachi–Hyderabad (intercity) Super Highway.

Like intercity road development, urban road development was supported by similar kinds of economic storylines. It was widely publicized that the Ring Road would have a substantial positive impact on the local economy and the development potential in Lahore. The storylines were based on changes in mobility pattern in the southern and south-western parts of Lahore in the form of 'greater levels of efficiency' and 'higher productivity'. However, Banister and Lichfield have raised many questions about such judgements concerning transport investment decisions.[60] They argued that new transport investment will bring changes in land values and rent, and result in a transfer of activity from one location in the city to another. The relative competitive position of one location will be enhanced, but at the expense of another location. This may also result in longer travel distances, more frequent trips, and an increase in car-based movements associated with low-density development. In the case of Lahore, new car movements have already become apparent and are replacing shorter trips by non-motorized transport in high-density areas near the northern and eastern sections of the Ring Road; this is resulting in higher demand for energy and a threat to the urban environment. Although it is widely accepted that the Ring Road will have physical, economic, social and environmental effects in Lahore, these effects were described only in general terms in all planning and detailed design reports. According to Hall and Banister, the precise relationships between transport infrastructure investments and different factors of urban development are not theoretically and empirically well known.[61] There seems to be no methodology to test the relationships and decisions that have been made according to belief rather than a solid understanding based on empirical evidence.

In the absence of appropriate financial capacity in local transport institutions, the storylines associated with financial gains such as 'road user charges' and 'parking charges' became popular in government organizations in 1990. These storylines were promoted to collect charges by way of additional tax on petrol. In this way, people travelling in a luxury car on a limited access highway or in poorly serviced public transport on low-grade roads had to pay an equal amount in road charges.

However, due to widespread resistance from the public who apprehended that it would raise the cost of public transport, these storylines did not gain enough strength to prevail. In fact, the government implemented a 'toll' tax for entering large cities and on intercity routes which ultimately raised public transport fares. The aims of these storylines were not to relieve congestion but to support financial gain. Ultimately the financial gains were used to build more bypasses, highways, motorways and ring roads.

In the early 1990s, the environmental storylines that emerged were associated with new problems of transport development. However, the change to the 'new realism' (Goodwin's 1996 phrase) has not occurred in Pakistan, and the old paradigm has always been supported by emerging environmental storylines. These storylines run around 'regulatory measures and technological fixes' to solve environmental problems caused by urban road transport. Such storylines contributed to the existing transport paradigm. All environmental studies argued that the environment could be improved by building more roads for the smooth flow of traffic by technical improvement of vehicles and by introducing environment-friendly fuel. Similar explanations were provided for air and noise pollution which it was said could be addressed by building high quality roads. Sometimes the issue of environment was deemed to have been addressed by 'landscaping measures' along roads and interchanges. In fact, these practices were advocated as a paradigm of sustainable development in the country. The origin of these beliefs lie in the World Bank document which states that 'expansion of transport infrastructure can reduce total pollution loads as congestion falls, average vehicle speeds rise and routes are shortened'.[62] Due to increase of pollution levels in urban areas, a 'window of opportunity' emerged to challenge existing thoughts about the consequences of transport policy. A similar opportunity has appeared in the rising price of oil which is moving beyond the purchasing power of the public. However, oil prices were dealt with by the introduction of CNG and LPG vehicles which are cheaper to run and are environment-friendly.

Current investigations have documented a notable shift in the storylines for road construction during the last fifty years. The shift ranges from economic growth to poverty reduction, and finally environmental sustainability—in particular the increasing air pollution. All these storylines have given priority to short term economic and environmental concerns while ignoring the ecological foundation of urban transport planning. In fact, the discursive journeys of these storylines have provided stability for the existing urban transport policy in Lahore. The planners must understand the way these storylines work in combination to stabilize policy and to resist institutional change.

Hajer argues that for a discourse to assume dominance, structuring and institutionalization of discourse must occur.[63] His arguments show that wider organizational, regulatory and legal arrangements are necessary for the acceptance of popular storylines. In Pakistan, storylines are developed within economics and environmental frames at the inter-urban level; and within town planning, engineering, economy and environmental frames in urban transport planning in Lahore. The process of discourse structuring can be observed by revealing the organizational forms of transport planning in Pakistan and in particular in Lahore. A major change towards sustainable transport has not been observed in different departments of the national, provincial and local institutions for transport planning. However, the creation of the NHA at the federal level and a series of transport policies in the decade of the 1990s underlined the government's greater recognition of transport. Also, environmental protection departments at the national and provincial levels were created to signal a commitment to the environment. The approval of the NCS (Agenda 21) in 1992, creation of the Pakistan Environmental Protection Council and agencies and enactment of the Pakistan Environmental Protection Act 1997, highlighted the government's recognition of the need to do things differently in relation to the environment. However, after more than a decade, all organizational changes proved to be merely strengthening the road development paradigm by accommodating the above mentioned storylines. The creation of environmental protection agencies and laws had no demonstrable effect on the current practice of road development.

Organizational stability was a key feature in the city of Lahore. In the Punjab, the C&W not only became the primary institution for road building but was also involved in planning and implementing the Lahore Ring Road project which was previously the responsibility of TEPA in LCDG. In contrast, the Punjab transport department and TEPA, both responsible for urban transport planning in general and public transport planning in particular, were weakened over time. However, evidence from the franchised public transport policy in Lahore suggests that these institutions tried to introduce new concepts in public transport planning. Regardless of how one views the policy analysis of the franchise system, this initiative did highlight the problem of public transport in Lahore. However, national and international institutions took no notice and instead worked closely with the C&W department in building the Lahore Ring Road and increasing capacity for the maintenance and management of other roads in the city. In parallel with the construction of the Ring Road, many storylines were developed and propagated as discussed earlier.

Healey suggests that discourses operate at three levels: at the 'surface level' of discussions; at the level of a 'system of meaning'; and at the 'deeper cultural level'.[64] This book has shown that the discourse on road building diffused from government organizations to research and community institutions. Even the public transport operators who were interviewed saw road building as being supportive of public transport operation. By and large, people have accepted these storylines. The frame of reference of experts from international, national and local institutions had fixed viewpoints that resisted change and acceptance of different options. Recently some alternative views were presented in local newspapers and conferences, but there is a need to develop research in the local universities. The implications of the alternative view—'sustainable transport'—have not yet been recognized. Increasing pollution and the rising price of oil has prompted change in the way people think and act, but transport experts are unwilling to stop further road building projects and will not work to improve public transport. Due to a continuous decline in the quality and services of public transport, individuals are moving towards private cars. In fact, people consider a car as a necessary adjunct to a better lifestyle and status. In the absence of appropriate funding for public and non-motorized transport, and lack of coordinated land-use planning, people's travel patterns have shown a movement away from sustainable transport. This indicates that we are not valuing or developing storylines to support the virtue of our large number of public transport and non-motorized trips in urban areas. To the contrary, these trips have become a symbol of poverty. Until subsidies for the marginal car-based modes are not revoked, it is more likely that the road building paradigm will be strengthened in future.

To conclude, the storylines supporting road building have been largely diffused in the practices and mindsets of transport policy makers, both local and those from international institutions working in Pakistan. This diffusion is particularly found in road development institutions. Perhaps these storylines are widely accepted by the public as well. Indeed, to displace these storylines even with the passage of time appears to be hard in Pakistan. The ambiguity of certain storylines, especially around employment opportunities, poverty reduction and environmental concerns was manipulated to produce policy solutions that in reality moved against sustainability by building more and more roads. The fundamental need now is for research to find ways to remake and reconstitute discourse. Storylines for improvement of public transport also need to co-exist with road building storylines. While all transport policies are now committed to environmental values and to employment generation, much greater investment was provided for road development than for collective and active transport.

In short, all three forms of path dependence—institutional relationships, techniques and storylines—helped one another to strengthen the road building paradigm. What is less clear is which has more dominance over others to create path dependence. But it is clear that just changing the techniques or creating another institution will not change the path-dependent nature of transport policy. We have to go beyond these concerns towards discursive change as well.

THE IMPORTANCE OF A PATH DEPENDENCE FRAMEWORK

Does the concept of path dependence provide a useful way of understanding and elucidating policy change and stability in current analysis? This section summarizes the previous discussion associated with the path dependence concept in policy analysis. The tools of path dependence, i.e., critical juncture (when events combine to develop a particular path) and self-reinforcement mechanism (when different forces stabilize the path according to the selection developed early) have been used to organize the argument.

The concept of 'critical juncture' in history offers a useful proposition for explaining the momentum of current policies. Beside historical evolution, a focus on 'critical juncture' helps to identify a precise time period, and mapping small events responsible for embedding a particular policy in an institutional context. The analysis showed the nature of the relationship between international and local institutions in shaping transport policy outcomes in a specific time frame when their future consequences were unclear. The choice of policy was not determined according to suitability but as a result of small events at a particular time which pushed transport development into the roads paradigm resulting in the elimination of environmentally superior transport options such as railway, public and non-motorized transport. Like technologies, policy options not chosen during the critical juncture might well disappear.[65]

However, the analysis also showed that alternative modes of transport still exist and have not completely disappeared, although the allocation of resources in favour of the selected (roads) policy affects their performance.

Mahoney alludes to the danger in path dependence analysis in simply tracing outcomes in history without explaining the occurrence of a specific outcome.[66] However, my analysis has reasonably explored a particular 'critical juncture' in history and each critical juncture may be different in different locations. Under the historical juncture, the evolution of a certain policy reflects preferences which may transform policies, but not necessarily, as discussed before in the case of Singapore. However, in Pakistan the historical juncture showed root effects and became a mere carrier. It would seem that history becomes important but at the same time there are limits to the predictability of the policy future. There is always the chance that during a reinforcing period, because of the capacity of local institutions, the outcomes are different from those that had tended to emerge from the critical juncture. There are two other possibilities in a policy process—the absence of critical juncture or there being more than one critical juncture—which may affect the shape of the policy and its outcome. However, in my analysis, the critical juncture limited the emergence of relationships between international and local institutions. It is possible that this critical juncture was either deliberately delivered or occurred unintentionally. The absence of strong local institutions and strong belief in roads as a tool for economic development provides a clear explanation for the lack of resistance during the critical juncture and the recognition of the strength and perception of international institutions. To sum up, critical juncture helps to explain how current policies can acquire the characteristics of the past. There is no question that the international development institutions' involvement in the mid-1950s and early 1960s left a strong imprint on the transport development of Pakistan. However, my analysis shows that the critical juncture 'happened and never un-happened' (using David's 2001 term).

The current study has explained the self-reinforcement mechanism in three fields: institutions, techniques and discourse, each responsible for stabilizing the path enacted during the critical juncture. The extension of the concept of path dependence in the institutional field supports the exploration of propositions in which the current transport policy is rooted. The explanation of the transport decision-making process clearly demonstrated how international institutions encouraged local institutions to choose and maintain one form of transport policy over time and control transport issues by identifying problems in a specific way. These international institutions have promoted just one path of development, that of the 'Western' nations. As their solutions take the form of large investment projects, local institutions have to rely on international institutions to provide expertise thus enabling the latter to control the decision-making process. This control helps to incorporate even environmental debates into the existing transport policy paradigm. The transfer of this particular path from the developed world results in a mismatch of transport policy with local needs for mobility and safety as well as ecological sustainability. The study has found that institutional path dependence provides a reasonable explanation for this mismatch. The use of the institutional path dependence concept helps to bridge the theoretical gap between the generic and the specific. It reveals that context is important because it comprises events in history and the components that shaped overall policy outcomes. In fact, these contexts are the source of differences amongst several policies in varied areas. The dysfunctional path of transport planning can only be avoided by building the capacity of local institutions to analyse transport problems from local perspectives. Overall, the institutional field in the path dependence approach explains the decision-making process and improves our understanding of the policy process.

The extension of the concept of path dependence in the technical field helps to trace the technical justification of the ongoing transport planning paradigm. The current study found technical path dependence in transport planning because techniques used in Pakistan were similar to those used elsewhere in the world. Although the empirical evidence affirms Dimitriou's work in urban transport planning, the current study differs from his findings in that the technology transfer strategy is more harmful than the technique itself.[67] The current analysis found that techniques carry weight due to their scientific appearance. However, there is a need to develop similar kinds of techniques for sustainable urban transport.

The focus on discourse in policy analysis helps to trace the movement of thoughts responsible for reinforcing the existing path in transport policy and planning. The analysis enables one to pay attention to the use of language and practice in policy analysis. The dominance of discourse which is typically comprised of storylines can thus be a key feature in explaining the dominant policy paradigm. In fact, storylines identify and help to make evident the relations amongst different organizations. The current analysis shows that the use of specific language has an effect on power relations amongst different institutions and the language is then further shaped by those very power relations. Additionally, the combination of institutions and discourse not only favour a particular policy but also prevent the adoption of an alternative vision of transport planning. There is one criticism of employing discourse in policy analysis which is that 'policy texts are usually written for a particular audience'.[68] The current analysis shows that discourse fields help to find relationships between language and power in shaping a particular policy discourse in transport planning. The discourse helped the particular path of road construction to penetrate to a deeper cultural level resulting in a perception of the problem that is similar both in the community at large and amongst experts. The discourse in the form of storylines created the belief which persuaded people to see economic and transport problem in terms of a dearth of roads. The study has concentrated on the text used by international and local institutions to develop policy documents, and newspaper stories, for promoting a particular version of transport solutions. Although Vigar saw the number of alternative storylines as a healthy situation, the present study did not find any strong alternative storylines.[69] In conclusion, discourse emerged as one factor of self-reinforcement in path dependence which helped to stabilize the path in the long run. Therefore, without addressing deep-seated beliefs, a lasting institutional change cannot be achieved.

This book has attempted to look at the concept of path dependence in a sustainable urban development perspective by examining the policy process that evolves during critical juncture. The analysis affirms the stability of policy enacted during a critical juncture. The reasons for the dominance of a particular policy path are historical, but it is dangerous to put full responsibility on a single factor. In fact, this responsibility must also be shared by factors such as lack of professional, financial and human resource capacity in local institutions and the influence of international institutions. The study showed how institutions, techniques and discourse fields made the policy gradually locked-in to an outcome which did not match sustainable options. But at the same time the situation is not easy to alter. The recognition of all three types of fields helps to illustrate how the self-reinforcement mechanism works subsequent to the critical juncture. The evident linkages among the three types of fields in path dependence define the characteristics of barriers which need to be overcome for achieving sustainable urban transport. Overall, the framework provided by the concept of path dependence has proved useful. The focus on different fields provides a way of looking at path dependence that is different to that in political science and economics. However, the discussion in these disciplines helps to generate a better understanding of the concept applied usefully in urban policy analysis. In short, path dependence is more than policy or institutional evolution based on a critical juncture in history. The concept associated with three different types of self-reinforcement mechanisms is responsible for the stability of a particular policy.

NOTES

1. Mitchell and Rapkin 1954.
2. Flyvbjerg et al. 2005, Pickrell 1992 and Richmond 1990.
3. SACTRA 1994.
4. Hall and Banister 1995: 284.
5. Thorner 1955.
6. In economics, the rate of return on investment refers to the benefits to an investor relative to the cost of the initial investment. The simplest way to measure the rate of return on investment is to measure the internal rate of return, i.e. the discount rate that applies to the discounted flow of costs and benefits of the project adding up to zero. It is the interest rate used to discount future net incomes that gives a net present value equal to zero. http://en.wikipedia.org/wiki/Rate_of_return_on_investment
7. World Bank 1996 cited by Dimitriou 1998: 342.
8. LDA 1997: 6–38.
9. Ibid.
10. Ibid.
11. TEPA and JICA 1992: 10–2.
12. Badami 2003: 23.
13. TEPA 1993: 76.
14. Barter et al. 2003.
15. Newman and Kenworthy 1999: 192.
16. Ibid.
17. Transport Expert, World Bank.
18. Tiwari 2002: 98; Newman and Kenworthy 1999: 162.
19. Whitelegg 1993.
20. Dimitriou 1998: 333.
21. Ibid.
22. Wilson et al. 1994.
23. JICA 1983, 1988; JICA and NTRC 1994.
24. Platteau 2004.
25. Sohail et al. 2006.
26. LDA 1997.
27. Mees 2000.
28. LDA 1997.
29. Tiwari 2002; Pucher et al. 2004.
30. Tiwari 2002.
31. Pucher et al. 2004.

32. Vasconcellos 2001.
33. Cervero 1998; Mees 2005.
34. Kah 2001.
35. Haider and Badami 2007.
36. Banister 2002: 215.
37. http://www.statpak.gov.pk/depts/pco/statistics/demographic_indicators98/demographic_indicators.html
38. 1898; 1946; 1985.
39. Hall 2000: 29–39.
40. Sussman 1976.
41. Hall 2000.
42. LDA 1997.
43. Banister 2002: 22.
44. Vigar 2002: 165.
45. Banister 2002: 131.
46. Banister 2002: 134.
47. Naess 2006.
48. Flyvbjerg et al. 2005.
49. Naess 2006.
50. National Audit Office 1988, para 4.1 cited by Banister 2002: 136.
51. SACTRA 1994.
52. Morisugi and Hayashi 2000.
53. Dimitriou 1995: 124.
54. Eade 2000.
55. Dimitriou 1995: 123.
56. Dimitriou 1992.
57. Dimitriou 1995: 125.
58. Innes 1995; Healey 1997; Vigar 2002 and Wilson 2001.
59. Banister 2002: 208.
60. Lichfield 1995: 4.
61. Banister 1995.
62. World Bank 1994: 22.
63. Hajer 1995.
64. Healey 1999.
65. Kasarda 1985: 39.
66. Mahoney 2000.
67. Dimitriou 1992.
68. Jacobs 2006: 47.
69. Vigar 2002: 211.

9

Conclusion: The Way Forward

The aim of the book was to identify and examine institutional barriers that create a hindrance to sustainable development, and prevent the application and implementation of sustainable urban transport policies in Pakistan. Accordingly, institutional barriers in the form of path dependence and its sub-fields: institutions, techniques and discourse, were identified and explained in transport planning of Pakistan. The analysis showed that sustainable transport cannot be achieved without addressing these institutional barriers. The essence of the concept of path dependence is that once a policy path has been taken, subsequent events and institutional arrangements strengthen the path and obstruct change in policy. Transport planning and policy in Pakistan clearly show that the road development model was adopted due to intervention of international institutions to achieve economic, social and environmental goals. Local institutions had insufficient professional capacity to make use of this intervention for promoting existing sustainable transport modes in Pakistani cities.

Despite this situation, local institutions in Lahore initially attempted to address cycling, walking and other non-motorized transport. Eventually, local institutions became uncertain about these modes and fully 'locked-in' to the road development paradigm. In fact, evidence from interview participants and policy documents suggests a very weak and vague thought process on the development of sustainable transport in Pakistani cities. This large scale gap of awareness is due to a complex range of path dependencies that arise as barriers to the development of a sustainable transport system in Pakistan. The book concludes that in order to understand such debates one needs to develop knowledge of how this situation was arrived at, and how such knowledge can contribute in shedding light on the future possibility of institutional change. In summary, this book makes three main conclusions:

1. Institutional path dependence was found in the local institutions for the development of sustainable transport planning in Pakistan. The analysis highlights the concept of 'critical juncture' and 'self-reinforcement mechanism' in path dependence. These concepts help us to identify institutional factors that changed the transport planning direction from railway and non-motorized transport to roads, and ultimately helped to stabilize planning in this direction. This change was made possible due to large-scale financial and technical assistance available from international institutions. While the role of international institutions was critical in advancing the road development paradigm, the analysis has particularly identified how weak institutional capacities of national institutions in general and local ones in particular had a significant impact in promoting the road-based transport paradigm. In Pakistan, transport planning has traditionally been divided between road and public transport organizations. But, while road organizations are strengthened by a hierarchy that extends from the local to provincial and federal government levels, public transport organizations in contrast mainly exist at the provincial level. There is practically no mechanism to ensure that sustainable development will be incorporated in all transport related development. In fact, there is a serious deficiency of professional capacity amongst local institutions. A sophisticated institutional culture is lacking due to low financial capacity, shortage of human resource, and a lack of skills, knowledge and research necessary to facilitate the process towards sustainability. It has been found that transport remains a highly centralized policy sector due to continuing dependence of local institutions on provincial and national institutions, and ultimately on the finance of international institutions. For these reasons, transport policy became institutionally path dependent on the rules, norms, expertise and organization of international institutions. Therefore, a strong and unambiguous commitment for sustainable transport from international institutions and a clear policy statement and capacity building programme from the federal government is essential for promoting sustainable transport at the local level.
2. The analysis highlighted the significance of techniques and methods for maintaining the stability of the road development paradigm in Pakistan. In fact, transport planning in Pakistan locked-in on the 'decide and provide' and 'predict and provide' techniques. There is a big gap between these techniques and institutional capacities

which make the transport planning process rather irrelevant for Pakistani cities. In fact, transport plans provided a technical base for institutionally pre-determined decisions and the initiative of the Lahore Ring Road in the early 1960s affirms this claim. These techniques therefore are barriers for the development of sustainable urban transport planning in Pakistan. Derived from experience in the developed world these techniques have become part of the technology transfer process from developed to developing countries; they are limited in scope and fail to develop a range of solutions necessary for Pakistani cities. The analysis concluded that the technical form of path dependence is creating a barrier to policy change and innovative thinking to address the long-range goal of sustainable development.

3. Finally, the analysis shows the importance of discourse in long term stability of the road development paradigm in Pakistan. The approach based on discursive path dependence, by identifying storylines, provided a framework for looking at the policy statements and planning documents to highlight how path dependence can be understood and how change can be managed in future.

   The discursive path dependence in transport policies in Pakistan was made possible due to various storylines. The study has documented a notable shift in the storylines for the road development paradigm during the last fifty years. The shift ranges from economic growth to poverty reduction and finally environmental sustainability, in particular increasing air pollution. All these storylines have given priority to short term economic and environmental concerns while ignoring the ecological foundation of urban transport planning. In fact, these storylines promote the view that better land-use planning, efficient transport links, economic and employment growth and environmental sustainability are only possible if an adequate transport infrastructure is provided.

Therefore, a new 'critical juncture' has not occurred in transport planning of Pakistan in spite of the debate on sustainable development. This is due to the belief of decision-makers in what they understand to be the right and true solution for transport in Pakistan. The analysis showed that we need a better understanding of how officials act on their beliefs, and where these beliefs come from. The different storylines show that most beliefs came from the professional understanding embedded in respective institutions. The study showed that these storylines work together for the stability of the old policy, and become resistant to institutional change. Therefore, the study concluded that language advances certain ideas, prompting certain values, defining certain roles and eventually proposing certain outcomes which shape the overall knowledge. These patterns of belief are ultimately hard to change.

In summary, this book concludes that Pakistan's transport development has become institutionally, technically and discursively path dependent upon international development institutions and their mental model developed according to the Western traditional urban transport planning knowledge. All three forms of path dependence are reinforcing one another to stabilize the policy enacted during the critical juncture. Therefore, all three forms of path dependence are equally important and require full treatment for institutional change. Together these points suggest that the consequences of road-based transport development for ecological conditions, future energy crisis and large number of road crashes has failed to bring a major shift in the approach of transport planning in Pakistan. In fact, the full range of environmental issues continuously arising from road development is discarded by adopting short range techniques and the dominant old frame of reference. Thus, sustainable development does not drive transport planning in Pakistan. It seems that sustainability is not the focus of our policies but is just given lip service.

In fact, all three aspects of path dependence are creating major barriers for sustainable urban transport planning in Pakistani cities. What is less clear is which aspect has more dominance than others in forming path dependence and how big is the difference amongst them in terms of their influence. But it is clear that just changing the policies or simply formulating sustainable urban transport policies, techniques and changing organizations are not sufficient to change the nature of transport policy. Institutional change should cover more than just organizational structure, which has happened many times in history. These institutional changes should include local institutional capacity building, change of mental models, change of assumptions and techniques and change of social and financial priorities.

The conclusion on institutional, technical and discursive path dependence has helped us to understand how institutions create barriers for sustainable urban transport, but also provide the opportunity for managing change over a long span of time. The study clearly shows that an understanding is needed of what stops true sustainable transport, along with what should be done in future. Transport planners cannot rely on another critical juncture (due to external shock) but should also try to create deliberate institutional change by analysing their own policies in a historical perspective. To tackle such issues suggests that an agenda of sustainable transport needs to be recognized at the national level with the creation of a 'National Transit Authority' (NTA). The NTA should be responsible for the development of strategic guidelines, policy packages and funding mechanisms in favour of sustainable transport. These policies may include the development of comfortable, frequent, reliable and inexpensive public transport, promotion of non-motorized transport, reallocation of road spaces and integration of transport and land use. These programmes must be based on quality research and new techniques developed at the local universities and research centres. This research and techniques should be developed keeping in mind the social, environmental and economic benefits of reduction in travel and ultimately traffic from existing cities. Overall, the federal government should take responsibility for the development of strategic goals for urban transport in Pakistan. The setting up of an NTA requires a great deal of political courage based on strong capacity building programmes in public, academic and community institutions at the national, provincial and local level.

The success of urban transport, especially in Lahore, requires strengthening of local institutional capacities so that they can implement a clear and transparent decision-making process, undertake skill development programmes, induct fresh staff and adopt a multidisciplinary approach to urban transport. Only strong and capable transport institutions can transform existing opportunities in Lahore such as high-density, low-car ownership and large number of non-motorized and public transport trips into the development of a sustainable transport system. Overall, Lahore does not need extra money to fix its urban transport problem. Lahore needs functional institutional arrangements with revised priorities in accordance with a sustainability criteria. These institutional arrangements should work for the integration of transport planning with the overall urban development vision of the city. Although fixing institutions is less exciting, of the current study clearly shows that it is the essence for the development of sustainable transport in Pakistan. A substantial amount of knowledge has already developed to support the formulation of sustainable transport policies, while this book contributes towards understanding the institutional dimension necessary to make change in favour of sustainable transport.

LIMITATION OF ANALYSIS

A number of factors limit generalization of the study to other areas and situations. These factors include: (1) an analysis which focused on specific contextual knowledge, (2) the application of theory and practice in a specific context that resulted in the exclusion of other notable knowledge, (3) methodology and data limitations, and (4) the need for a more detailed institutional analysis.

The study focused on a review of literature in the areas of transport and sustainable development in conjunction with public policy and institutionalism. In terms of theory, the analysis focused on the notion of path dependence and tried to identify its key components. These areas of knowledge are extensive and required considerable time to undertake a comprehensive review. Therefore, other important areas of knowledge such as political economy and selective planning theories, which would have improved the analysis, were not examined in depth.

The selection of one case study in a single country also results in some limitations. There is a possibility of different kinds of path dependence in other transport projects, cities and even countries, where international institutions have been involved. However, one case study in one city of Pakistan is reasonably representative of this country for the purpose of the current analysis. For the selection of the case study, all major cities and their urban transport projects proposed in various studies were taken into account, and criteria such as representation in national policy making, scale of the projects, involvement of international institutions and government perception of economic, social and environmental sustainability were taken into account. However, the influence and power of the motor car and both local and international road construction industries, as also professional groups, in ensuring the stability of road-based transport policies were not considered and require separate attention. Moreover, the study did not examine fully the role of Lahore and the Lahore Ring Road in its regional context. The complex nature of transport planning and interrelationships of many institutions at the regional level are important considerations in studying sustainability of urban transport in a metropolitan city.

In addition to methodological limitations, there is an evident shortage of data. However, the author collected sufficient data which included all major policy and planning documents issued over many years to have a broader base for analysis. The author had the advantage of having previous transport planning experience in Pakistan, and his knowledge of the institutional system, language, culture and relevant contact persons enabled him to get insights, information and data. However, small reports, letters, detailed financial documents and in-house discussion papers were hard to access due to the culture of confidentiality in Pakistan. Similarly, an effort had been made to interview a representative sample of persons and institutions involved in transport planning in Pakistan and Lahore. However, requests for interviews with senior officials of local and international institutions were consistently answered by junior staff with limited authority and experience in these projects. Additionally, interviews were conducted in three different languages, i.e., English, Urdu and Punjabi and permission to make audio recordings was not given. There is the possibility then that in interview citations potentially significant thoughts, ideas and insights might have been missed. However, care has been taken to maximise information and to minimize misinterpretation of important ideas. The study acknowledges the difficult accessibility of data for tracing the flow of policy and allocation of funds from international to local institutions in Pakistan. Therefore, this study recommends that a continuous effort be made to create an open planning process and free access to institutional data.

A qualitative analysis has been employed to identify the nature of path dependence in transport planning of Pakistan but this process might be limited because it involves interpretation. The study did not employ any large scale quantitative analysis because it required extensive data collection. However, the book has made a small contribution to a larger body of theory and empirical research by identifying three forms of path dependence in transport planning of Pakistan. However, care should be taken in generalizing its findings to other developing countries where, like Pakistan, the role of international institutions is dominant.

THE WAY FORWARD

This book clearly shows that sustainable development cannot be achieved without addressing the institutional aspect of urban transport planning in Pakistan. The analysis employed the concept of 'path dependence' to address the institutional dimensions and barriers in Pakistan. Pierson identifies two distinct steps in exploring path dependence in institutional analysis.[1] The first step is the analysis of policies over time, identifying the barriers which limit the ability to implement institutional change. The current study has identified institutional, technical and discursive forms of path dependence as barriers to sustainable urban transport planning in

Pakistan. (1) In institutional path dependence, while the position of international institutions was critical in advancing the road development paradigm, the current study has particularly identified how weak institutional capacities at provincial and local levels were significant in promoting the road-based transport paradigm. (2) The study found technical 'lock-in' as a barrier to the development of the range of solutions necessary for Pakistani cities. (3) The study showed the importance of conventional discourse or mental maps underpinning existing policy settings. These three forms of path dependence reinforce one another over time and need to be exposed in order to advance to the next step.

Pierson's second step seeks opportunities for the development of important propositions about the process of institutional development.[2] However, the process of institutional development cannot be expected to occur simply as a result of change in policies, political will and spontaneous restructuring of organizations, though these are important. Rather, institutional development is most likely to occur by developing the capacity of local institutions to plan and manage urban transport. All three types of path dependence can be made open to question and perhaps dislocated in favour of more sustainable options.

Therefore, a future study will examine in detail the capacity of local institutions in the city of Lahore. In the past, many studies conducted by the World Bank and their associated consultants proposed to build the capacity of local institutions such as the creation of a traffic engineering and transport planning unit (TEPU) in LDA and a road management unit in the C&W department of the Punjab government. However, the main focus of these capacities was to increase the use of personal motorized vehicles supported by large-scale road development programmes. In contrast, institutional capacities for sustainable urban transport require an entirely different focus. The future study will explore where and what kind of capacity is required to promote sustainable urban transport. These capacities include the type of organization needed for sustainable transport, the development of appropriate transport decision-making processes, rearrangement of the structure of existing organizations, training of personnel and the training required to develop appropriate knowledge and skills.

Future rising oil prices and increasing dependence on oil in Pakistan may provide coverage to form a new 'critical juncture' stimulating institutional change. Oil imports in Pakistan increased from 54.5 per cent of the country's total requirement in 1991 to 71.8 per cent in 2004 while local production of oil will start to decline from 2010.[3] Moreover, the

World Bank financed 'Public Sector Capacity Building Project' and concerns arising on energy and public transport development in a newly released Pakistan 'Vision 2030' suggest that the time is ripe to develop propositions for the development of institutions for sustainable urban transport.

Social constructive institutionalists have placed a heavy responsibility on the catalytic role of previous 'losers',[4] and 'skilled social actors'[5] for institutional change. Pierson argues that marginal actors can bring partial or complete institutional change by creating pressure on existing institutions.[6] Thelen believes that 'losers' have the potential to bring institutional change because they incur less cost in changing the pattern of current institutions.[7] The future study sees these analyses as an opportunity for building social capital by entering into debates about civil society and reclaiming public life. It is also now being used by the World Bank with regard to economic and societal development and by management experts as a way of thinking about organizational development. However, caution should be exercised in transferring the concept to the developing world, and the possibility of 'elite capture' should not be ignored.[8] Therefore, the future study will explore the detailed features of social capital building necessary for institutional development in the context of Pakistan.

Current analysis found that academics and civil society as well as the weak 'public transport' lobbies had no role in transport planning and decision making, and noted the absence of a 'non-motorized transport' organization from the process. In the presence of a large proportion (75 per cent) of trips made by public transport and non-motorized transport, there is an opportunity to develop propositions to organize and strengthen the above mentioned 'losers' and incorporate them into transport decision-making. In the case of Australia, there are relatively strong pro public transport lobbies (such as the Public Transport Users Association in Victoria). The creation of a public transport corporation in Victoria in the 1980s contained the potential to break the path of road building dominance in transport policy. But it did not last long enough to make a lasting impact. In Perth (Western Australia) a similar organization has had much greater impact, as has the integration of transport and land-use planning under the department of planning and infrastructure of Western Australia. In Pakistan, the debate on global warming, high oil prices and the increasing number of pedestrians who are victims of road crashes can provide the opportunity to organize these marginal groups and transport modes and incorporate them into the institutional context. These marginal groups may work within the old institutions to bring about institutional change in urban transport planning.

Finally, the above objectives will be framed by the perspective that through the development of local research and techniques along with shaping 'mental models', an alternative model of sustainable transport can be achieved.

NOTES

1. Pierson 2004: 133.
2. Ibid.
3. *Economic Survey* 2004–5.
4. Clemens 1997; Thelen 2003.
5. Sweet et al. 2001.
6. Pierson 2004: 137.
7. Thelen 2004.
8. Platteau 2004.

References

Adams, B., 1993, Sustainable Development and the Greening of Development Theory, in, Schuurman, F. (Eds.), *Beyond the Impasse: New Directions in Development Theory*, London: Zed Book Ltd., pp. 207–22.

Aina, T., 1990, The Politics of Sustainable Third World Urban Development, in, Cadman, D. and Payne, G. (Eds.), *The Living City: Towards a Sustainable Future*, London: Routledge.

Arif, F., 2002, Transport Planning in Jakarta, *Urban Policy and Research*, 20(4), pp. 371–89.

Arthur, B., 1987, Path-dependent Processes and the Emergence of Macro-Structure, *European Journal of Operational Research*, 30, pp. 294–303.

Arthur, B., 1988a, Self-Reinforcing Mechanisms in Economics, in, Anderson, P., Arrow, K. and Pines, D. (Eds.), *The Economy as an Evolving Complex System*, Redwood City: Addison Wesley Publications, pp. 9–31.

Arthur, B., 1988b, Urban Systems and Historical Path Dependence, in, Ausubel, J. and Herman, R. (Eds.), *Cities and their Vital Systems: Infrastructure, Past, Present and Future*, Washington: National Academy Press, pp. 85–97.

Arthur, B., 1988c, Competing Technologies: an Overview, in, Dosi, G. et al. (Eds.), *Technical Change and Economic Theory*, London: Printer Publisher, pp. 590–607.

Arthur, B., 1989, Competing Technologies, Increasing Returns and Lock-in by Historical Events, *The Economic Journal*, 99(1), pp. 116–31.

Arthur, B., 1994, *Increasing Returns and Path Dependence in the Economy*, USA: The University of Michigan Press.

Arthur, B., Ermoliev, Y. and Kaniovski, Y., 1983, On Generalised Urn Schemes of the Poly Kind, *Cybernetics*, 19, pp. 61–71.

Asian Development Bank (ADB), 1993, *Pakistan Urban Sector Profile*, Philippines: Asian Development Bank.

Asian Development Bank, 1998, *Asia Least-cost Greenhouse Gas Abatement Strategy (ALGAS) Pakistan*, Manila: Office of Environment and Social Development, Asian Development Bank.

Asian Development Bank, 2005, *Country Strategy and Programme Update 2006–08—Pakistan*, Manila: Asian Development Bank.

Asian Development Bank and Government of Pakistan, National Transport Research Centre, 2003, *National Transport Policy*, Islamabad.

Badami, M., 2003, Urban Transport Planning for Sustainability and Justice in Less Industrialised Countries: The Case of India, Paper Presented at the AESOP–ACSP Joint Congress, Leuven, Belgium.

Banister, D. and Lichfield, N., 1995, The Key Issues in Transport and Urban Development, in, Banister D. (Eds.), *Transport and Urban Development*, London: Spon, pp. 1–16.

Banister, D., 2000, Sustainable Urban Development and Transport—A Eurovision for 2020, *Transport Reviews*, 20(1), pp. 113–30.

Banister, D., 2002, *Transport Planning*, London: Spon Press.

Barter, P., Kenworthy, J. and Laube, F., 2003, Lessons from Asia on Sustainable Urban Transport, in, Low, N. and Glesson, B. (Eds.), *Making Urban Transport Sustainable,* London: Palgrave, pp. 252–69.

Berman, S., 1998, Path Dependency and Political Action: Re-examining Responses to the Depression, *Comparative Politics*, 30(4), pp. 379–400.

Black, W., 2001, An Unpopular Essay on Transportation, *Journal of Transport Geography*, 9, pp. 1–11.

Brecher, I. and Abbas, S., 1972, *Foreign Aid and Industrial Development in Pakistan*, UK: Cambridge University Press.

Buhrs, T. and Aplin, G., 1999, Pathway Towards Sustainability: The Australian Approach, *Journal of Environmental Planning and Management*, 42(3), pp. 315–40.

Burki, S., 1992, The Management of Crisis, in, James, W. and Subroto, R. (Eds.), *Foundations of Pakistan's Political Economy: Towards an Agenda for the 1990s,* New Delhi: Sage Publication, pp. 116–44.

Burki, S., 2000, Changing Perception and Altered Reality: Emerging Economies in the 1990s, Washington: The World Bank.

Campbell, C. and Laherrere, J., 1998, The End of Cheap Oil, *Scientific American,* March Edition, pp. 78–83.

Campbell, C., 2003, The Peak of Oil: An Economic and Political Turning Point for the World in, Low, N. and Glesson, B. (Eds.), *Making Urban Transport Sustainable,* New York: Palgrave Macmillan, pp. 42–65.

Cervero, R., 1998, *The Transit Metropolis: A Global Inquiry,* Washington: Island Press.

Chartered Institute of Transport (CILT) Pakistan, 1998, *Transport Policy of Pakistan,* available on http://www.pakroadusers.com/tsdi/cittp.htm#preamble

Cities Alliance, 2006, *City Development Strategy Guidelines: Driving Urban Performance*, Washington D.C.

Clemens, E., 1997, *The People's Lobby: Organisational Innovation and the Rise of Interest Group Politics in the United States*, 1890–1925, Chicago: University of Chicago Press.

Cohen, L. and Manion, L., 1989, *Research Methods in Education*, London: Routledge.

Cohen, M., 1996, The Hypothesis of Urban Convergence: Are Cities in the North and South Becoming more alike in an Age of Globalisation?, in, Cohen, M., Ruble, B., Tulchin, J. and Garland, A. (Eds.), *Preparing for the Urban Future, Global Pressure and Local Forces*, Washington: The Woodrow Wilson Centre Press, pp. 25–38.

Collier, R., 1993, Combining Alternative Perspectives: Internal Trajectories versus External Influences as Explanations of Latin American Politics in the 1940s, *Comparative Politics*, 26(1), pp. 1–29.

Connor, R. and Dovers, S., 2002, *Institutional Change and Learning for Sustainable Development*, Working paper 2002/1, Centre for Resources and Environmental Studies, Australian National University.

Considine, M., 1994, *Public Policy: A Critical Approach*, Melbourne: Macmillan.

Daly, H., 1996, *Beyond Growth: The Economics of Sustainable Development*, Boston: Beacon Press.

David, P., 1985, Clio and the Economics of QWERTY, *The American Economic Review*, 75(2), pp. 332–37.

David, P., 1997, Path Dependence and the Quest for Historical Economics: One More Chorus on the Ballad of QWERTY, Oxford Economic and Social History Working Papers.

David, P., 2001, Path Dependence, its Critics and the Quest for 'Historical Economics', in, Garrouste, P. and Ioannides, S. (Eds.), *Evolution and Path Dependence in Economic Ideas: Past and Present*, UK: Edward Elgar Publisher, pp. 15–40.

Deffeyes, K., 2001, *Hubbert's Peak, the Impending World Oil Shortage*, New Jersey: Princeton University Press.

Denzau, T. and North, D., 1994, Shared Mental Models: Ideologies and Institutions, *Kyklos*, 47, pp. 3–31.

Department of Environment, Transport and Regions (DETR), 2000, *Transport 2010—the 10 Year Plan*, London: DETR.

DiMaggio, P. and Powell, W., 1983, The Iron Cage Revisited: Institutional Isomorphism and Collective Rationality in Organizational Fields, *American Sociological Review*, 48(2), pp. 147–60.

DiMaggio, P. and Powell, W., 1991a, Introduction, in, Powell, W. and DiMaggio, P. (Eds.), *The New Institutionalism in Organizational Analysis*, Chicago: Chicago University Press, pp. 1–38.

DiMaggio, P. and Powell, W., 1991b, The Iron Cage Revisited: Institutional Isomorphism and Collective Rationality in Organization Fields, in, Powell, W. and DiMaggio, P. (Eds.), *The New Institutionalism in Organizational Analysis*, Chicago: Chicago University Press, pp. 63–82.

Dimitriou, H., 1992, *Urban Transport Planning: a Development Approach*, London: Routledge.

Dimitriou, H., 1995, *A Developmental Approach to Urban Transport Planning: An Indonesian Illustration*, Aldershot: Avebury.

Dimitriou, H., 1998, Transport and Sustainable Urban Development: Providing a New Lead, in, Dimitriou, H. and Cook, A. (Eds.), *Land Use/Transport Planning in Hong Kong: The End of an Era*, Aldershot: Ashgate, pp. 323–62.

Dina, N., 1899, *The Two Friends: A Descriptive Story of the Lahore Life*, Lahore.

Dobson, A., 1998, *Justice and the Environment: Conceptions of Environmental Sustainability and Theories of Distributive Justice*, Oxford: Oxford University Press.

Dovers, S., 2001, *Institutions for Sustainability*, Centre for Resource and Environmental Studies, the Australian National University.

Eade, R., 2000, Issues in Environmental Impact Assessment: What has Scoresby Taught us?, *Urban Policy and Research*, 18(4), pp. 515–27.

Easterly, W., 2001, *The Political Economy of Growth without Development: A Case Study of Pakistan*, Paper for the Analytical Narratives of Growth Project, Kennedy School of Government, Harvard University.

Elander, I. and Montin, S., 1990, Decentralization and Control: Central-Local Government Relations in Sweden, *Policy and Politics*, 18(3), pp. 165–80.

Elander, I., 1991, Analysing Central-Local Government Relations in Different Systems: a Conceptual Framework and Some Empirical Illustrations, *Scandinavian Political Studies*, 14(1), pp. 31–58.

Elster, J., 1986, The Market and the Forum: Three Varieties of Political Theory, in, Elster, J. and Hylland, A. (Eds.), *Foundations of Social Choice Theory*, Cambridge: Cambridge University Press, pp. 103–32.

Emerson, R., 1962, Power-Dependence-Relations, *American Sociological Review*, 27, pp. 31–41.

Ercelan, A., 2004, *Aid from Tokyo to Islamabad via Washington and Manila: Who Decide What Happens to Whom in Pakistan?*, Paper presented in International Conference on 50 Years of Japan ODA, 6–8 October, Tokyo.

Fatima, M., 2003, A Challenge: Pakistan's External Dependence Peaks at Golden Jubilee, available on http://www.pakistaneconomist.com/database1/macro-economy/macro-economy-77.htm

Flyvbjerg, B., Holm, M. and Buhl, S., 2005, How (in) Accurate are Demand Forecasts in Public Works Project?: The Case of Transportation, *Journal of the American Planning Association*, 71(2), pp. 131–46.

Frantzeskakis, J., 1995, Configuration, Hierarchy and Spacing of the Urban Road Network in Islamabad, *Ekistics*, 62, pp. 236–41.

Gaventa, J., 1980, *Power and Powerlessness: Quiescence and Rebellion in an Appalachian Valley*, Oxford: Clarendon Press.

Glover, W., 1999, *Making Lahore Modern: Urban Form and Social Practices in Colonial Punjab, 1849–1920*, Unpublished PhD Thesis: University of California, Berkeley, USA.

Goldstone, J., 1998, Initial Conditions, General Laws, Path Dependence and Explanation in Historical Sociology, *The American Journal of Sociology*, 104(3), pp. 829–45.

Goodwin, P., 1996, Road Traffic Growth and the Dynamics of Sustainable Transport Policies, in, Cartledge, B. (Eds.), *Transport and the Environment*, Oxford: Oxford University Press, pp. 6–22.

Gorges, M., 2001, New Institutionalists Explanations for Institutional Change: A Note of Caution, *Politics*, 21(2), pp. 137–45.

Goulding, H., 1924, *Old Lahore, Reminiscences of a Resident*, Lahore: Sang-e-Meel Publications.

Government of Pakistan, National Planning Board, 1957, *The First Five-Year Plan (1955–60)*, Islamabad.

Government of Pakistan, Planning Commission, 1960, *The Second Five-Year Plan (1960–65)*, Islamabad.

Government of Pakistan, Planning Commission, 1964, *Outline of the Third Five-Year Plan (1965–70)*, Islamabad.

Government of Pakistan, Communication and Works Department 1965, *Master Plan for Highways*, Howard Needles, Tammen and Bergendoff Consultants.

Government of Pakistan, Planning Commission, 1970, *The Fourth Five-Year Plan (1970–75)*, Islamabad.

Government of Pakistan, Central Roads Organizations, Ministry of Communication, 1971, *Guide to Highway Design Standards and Specifications*, Islamabad.

Government of Pakistan, Planning Commission, 1978, *The Fifth Five-Year Plan (1978–83)*, Islamabad.

Government of Pakistan, Planning Commission, 1983, *Sixth Five-Year Plan (1983–88)*, Islamabad.

Government of Pakistan, Planning Commission, 1988, The Seventh Five-Year Plan (1988–93) and Perspective Plan (1988–2003), Islamabad.

Government of Pakistan, National Transport Research Centre (NTRC), 1991, draft *National Transport Policy*, Islamabad.

Government of Pakistan, National Transport Research Centre (NTRC), 1992, draft *National Transport Policy*, Islamabad.

Government of Pakistan, Ministry of Environment, Urban Affairs Division (EUAD) and IUCN, 1992, *The Pakistan National Conservation Strategy*, Islamabad.

Government of Pakistan, Planning Commission, 1994, *The Eight Five-Year Plan (1993–98)*, Islamabad.

Government of Pakistan, National Transport Research Centre, 1995, *Greater Islamabad/Rawalpindi Area Transportation Study—Final Report*, Islamabad: Scandiaconsult.

Government of Pakistan, Federal Bureau of Statistics, 1997a, *50 Years of Pakistan in Statistics, Vol. III*, Islamabad.

Government of Pakistan, Federal Bureau of Statistics, 1997b, *50 Years of Pakistan in Statistics, Vol. IV*, Islamabad.

Government of Pakistan, Planning and Development Division, 1997, *Manual for Development Projects: Preparation, Appraisal, Approval, Implementation, Monitoring and Evaluation*, Islamabad.

Government of Pakistan, Federal Bureau of Statistics, 1998, *Census Report*, Islamabad.

Government of Pakistan, Small and Medium Enterprise Development Authority (SMEDA), 1999, draft *National Transport Strategy*, Lahore.

Government of Pakistan, Planning Commission, 2000, draft *Transport Policy*, Islamabad.

Government of Pakistan, Ministry of Environment, Local Government and Rural Development, 2000, *A Hand Book on The Pakistan National Conservation Strategy*, Islamabad.

Government of Pakistan, Pakistan Environmental Protection Agency and Japan International Cooperation Agency (JICA), 2000, *3 Cities Investigation of Air and Water Quality: Lahore, Rawalpindi and Islamabad*, Islamabad.

Government of Pakistan, Planning Commission, 2001, *The Ten Year Perspective Development Plan 2001–11 and Three Year Development Programme 2001–04*, Islamabad.

Government of Pakistan, National Transport Research Centre (NTRC), 2001, draft *National Transport Policy*, Islamabad.

Government of Pakistan, Ministry of Environment, Local Government and Rural Development, IUCN and UNDP, 2001, *National Environmental Action Plan (NEAP)*, Islamabad.

Government of Pakistan, National Highway Authority, 2005, *Road Network Development*, Presentation made on Pakistan Development Forum 2005 at Islamabad.

Government of Pakistan, Planning Commission, 2005 *Working Draft Medium Term Development Framework*—MTDF (2005–10), Islamabad.

Government of Pakistan, Ministry of Finance, 2005, *Economic Survey 2004–2005*, Islamabad.

Government of Pakistan, Ministry of Finance, 2006, *Economic Survey 2005–06*, Islamabad.

Government of Pakistan, National Transport Research Centre (NTRC), 1992, *NTRC Bus Train*, Islamabad.

Government of Pakistan, National Transport Research Centre (NTRC), 1996, *The People's Train (Preliminary Evaluation)*, Islamabad.

Government of the Punjab, Housing and Physical Planning Department, 1973, *Master Plan for Greater Lahore*, Lahore: Master Plan Project Office.

Government of the Punjab, Transport Department, 1998, *Punjab Urban Transport Sector—A Promise of Profit and Community Service*, Lahore.

Government of Punjab, 2004, *Vision 2020—Pre Budget Policy Address of Ch. Pervaiz Elahi, Chief Minister*, Punjab, Lahore.

Government of Punjab, Communication & Works Department, 2004a, *PC1—Construction of Interchange at Niazi Chowk at Ring Road, Lahore*, Lahore: NESPAK.

Government of Punjab, Communication & Works Department, 2004b, *Lahore Ring Road—Construction of Road Portion from Niazi Chowk to Saggian Chowk, Lahore—Pre-Qualification Document*, Lahore.

Government of Punjab, 2005a, *Pre-Budget Address of Chief Minister*, Lahore.

Government of Punjab, 2005b, *Revised Estimates of Charged Expenditure and Demands for Grant (Development) for 2004–05*, Lahore.

Government of Punjab, Finance Department, 2006, *Medium Term Development Framework 2006–09 and Annual Development Programme 2006–07*, Lahore.

Griffiths, M., 2005, *Pushing the Energy Boundaries: 'Peak Oil' What is it and What Does it Mean for Planners?*, Paper presented in New Zealand Institute of Planners Congress 2005.

Grindle, M., 1996, *Challenging the State, Crisis and Innovation in Latin America and Africa*, Cambridge: Cambridge University Press.

Gusfield, J., 1981, *The Culture of Public Problems: Drinking-Driving and the Symbolic Order*, Chicago: University of Chicago Press.

Haider, M. and Badami M., 2007, *Balancing Efficiency and Equity in Public Transit in Pakistan*, in Laquian, A., Tewari, V. and Hanley L., Baltimore (Eds.), *The Inclusive City: Infrastructure and Public Services for the Urban Poor in Asia*, Baltimore: the Johns Hopkins University Press, pp. 122–143.

Hajer, M., 1995, *The Politics of Environmental Discourse: Ecological Modernization and the Policy Process*, Oxford: Clarendon Press.

Hall, P., 1986, *Governing the Economy: The Politics of State Intervention in Britain and France*, Cambridge: Polity.

Hall, P., 1993, Policy Paradigms, Social Learning and the State, *Comparative Politics*, 25(3), pp. 275–96.

Hall, P. ad Banister, D., 1995, Summary and Conclusion, in, Banister, D. (Eds.), *Transport and Urban Development,* London: Spon, pp. 278–87.

Hall, P. and Taylor, R., 1996, Political Science and the Three New Institutionalism, *Political Studies*, XLIV, pp. 936–57.

Hall, P., 2000, The Century of Modern Planning, in, Freestone R. (Eds.), *Urban Planning in a Changing World: The Twentieth Century Experience*, London: Spon, pp. 20–39.

Hamid, S., 2005, The Fashion of American Economic Rules in Pakistan, in the *Nawa-i-waqt*, 27 April (in Urdu language).

Hanson, A, Bass, S., Bouzher, A. and Samdani, G., 2000, Pakistan's National Conservation Strategy: Renewing Commitment to Action—Report of the Mid-Term Review, Islamabad.

Haq, M., 1973, Seven Sins of Economic Planners, in, Baqai, M. and Brecher, I. (Eds.), *Development Planning and Policy in Pakistan 1950–70*, Karachi: NISER, pp. 9–20.

Hartman, J. and Hedbolm, J., 1979, *Methods for the Social Sciences*, London: Greenwood Press.

Hasan, P., 1997, Learning from the Past: A Fifty-year Perspective on Pakistan's Development, *The Pakistan Development Review*, 36(4), pp. 335–402.

Hasan, P., 1998, *Pakistan's Economy at the Crossroads: Past Policies and Present Imperatives*, Karachi: Oxford University Press.

Hasan, S., 1999, *Pakistan: The Story Behinds its Economic Development*, Rawalpindi: Re-printed by Offset Printers.

Hasan, A., 2005, The Political and Institutional Blockages to Good Governance: the case of the Lyari Expressway in Karachi, *Environment and Urbanization*, 17(2), pp. 127–41.

Hasan, A. and Ali, Amenah, 1992, Environmental Problems in Pakistan; their Origins and Development and the Threats that they Pose to Sustainable Development, *Environment and Urbanization*, 4(8), pp. 8–21.

Headrick, D., 1981, The Railroads of India, in, Headrick, D. (Eds.), *The Tools of Empire: Technology and European Imperialism in the Nineteenth Century*, New York: Oxford University Press, pp. 180–91.

Healey, P., 1997, *Collaborative Planning: Shaping Places in Fragmented Societies*, Basingstoke: Macmillan.

Healey, P., 1999, Sites, Jobs and Portfolios: Economic Development Discourses and the Planning System, *Urban Studies*, 36(1), pp. 27–42.

Heinberg, R., 2003, *The Party's Over, Oil, War and the Fate of Industrial Societies*, Canada: New Society Publishers.

Heinberg, R., 2004, *Powerdown, Options and Actions for a Post-Carbon World*, Canada: New Society Publishers.

Hoberg, G., 1991, Sleeping with an Elephant: The American Influence on Canadian Environmental Regulation, *Journal of Public Policy*, 11, pp. 107–32.

Houghton J., Filho, L., Callander, B., Harris, N., Kettenberg, A. and Maskell, K. (Eds.), 1997, *Climate Change 1995: the Science of Climate Change*, New York: Cambridge University Press.

Howard, E., 1898, *To-Morrow: A Peaceful Path to Real Reform*, London: Swan Sonnenschein.

Howard, E., 1946, *Garden Cities of Tomorrow*, London: Faber and Faber.

Howard, E., 1985, *Garden Cities of Tomorrow*, New Illustrated Edition, with an introduction by Builth T. R., Wells: Attic Books.

Huda, M., 1968, Planning Experiences in Pakistan, *Pakistan Development Review*, 8, pp. 329–353.

Husain, I., 1984, Raising Resources for Development, in, Burki, S. and LaPorte. R. (Eds.), *Pakistan's Development Priorities: Choices for the Future*, Karachi: Oxford University Press, pp. 103–36.

Husain, I., 1999, *Pakistan the Economy of an Elitist State*, Karachi: Oxford University Press.

Immergut, E., 1998, The Theoretical Core of the New Institutionalism, *Politics and Society*, 26(1), pp. 5–34.

Imran, M. and Low, N., 2003, Time to Change the Old Paradigm: Promoting Sustainable Urban Transport in Lahore, Pakistan, *World Transport Policy and Practice*, 9(1), pp. 32–39.

Imran, M. and Low, N., 2005, Impact of Global Institutions on Urban Transport in Pakistani Cities, in, Greca, P. La. (Eds.), *Planning in a more Globalised and Competitive World*, Netherlands: ISoCaRP, pp. 253–64.

Innes, J., 1995, Planning Theory's Emerging Paradigm: Communicative Action and Interactive Practice, *Journal of Planning Education and Research*, 14(3), pp. 183–89.

Jackson, K., 1985, *Crabgrass Frontier: the Suburbanization of the United States*, New York: Oxford University Press.

Jacobs, M., 1994, Sustainable Development, *Journal of Developing Areas*, 28, pp. 237–52.

Jacobs, M., 1999, Sustainable Development as a Contested Concept, in, Dobson, A. (Eds.), *Fairness and Futurity: Essay on Environmental Sustainability and Social Justice*, Oxford: Oxford University Press, pp. 21–45.

Jacobs, K., 2006, Discourse Analysis and its Utility for Urban Policy Research, *Urban Policy and Research*, 24(1), pp. 39–52.

Japan International Cooperation Agency (JICA), 1983, *The Study on the National Transport Plan in the Islamic Republic of Pakistan*, Islamabad.

Japan International Cooperation Agency (JICA), 1988, *The Study on the National Transport Plan in the Islamic Republic of Pakistan*, Islamabad.

Japan International Cooperation Agency (JICA) and National Transport Research Centre, Government of Pakistan, 1994, *The Study on the National Transport Plan in the Islamic Republic of Pakistan*, Islamabad.

Jenkins, P. and Smith, H., 2001, An Institutional Approach to Analysis of State Capacity in Housing Systems in the Developing World: Case Studies in South Africa and Cost Rica, *Housing Studies*, 16(4), pp. 485–507.

Jessop, B., 1982, *The Capitalist State*, Oxford: Blackwell.

Jessop, B., 1990, *State Theory*, Cambridge: Polity Press.

Jessop, B., 1995, Towards a Shumpeterian Workfare Regime in Britain? Reflections on Regulation, Governance and Welfare State, *Environment and Planning A*, 27(11), pp. 1613–26.

Kah, J., 2001, Urban Transportation Privatization and Regulation in Sub-Saharan Africa: Problems, Policies and Prospects: Case Studies of Senegal and the Gambia, New Brunswick: Unpublished PhD dissertation at the State University of New Jersey, USA.

Kasarda, J., 1985, Urban Change and Minority Opportunities, in, Peterson, P., (Eds.), *The New Urban Reality*, Washington: Brookings Institutions, pp. 33–68.

Kenworthy, J., Laube, F., Newman, P., Barter, P., Raad, T., Poboon, C. and Guia, B., 1999, *An International Sourcebook of Automobile Dependence in Cities 1960–90*, Boulder: University Press of Colorado.

Khakee, A. and Low, N., 1996, Central-Local Relations in Sweden and Australia: Power, Responsibility and Territorial Equity in Comparative Perspective, *International Planning Studies*, 1(3), pp. 331–26p.

Khan, M., 1996, Problems and Prospects of Urban Environmental Management in Pakistan, *The Pakistan Development Review*, 35(4), pp. 507–23.

Kiser, L. and Ostrom, E., 1982, The Three Worlds of Action: A Metatheoretical Synthesis of Institutional Approach, in, Ostrom, E. (Eds.), *Strategies of Political Inquiry*, Beverly Hills: Sage Publication, pp. 179–222.

Lahore Development Authority (LDA) and World Bank/IDA, 1980a, *Lahore Urban Development and Traffic Study, Vol. 1-A—Urban Planning*, Lahore: LDA.

Lahore Development Authority (LDA) and World Bank/IDA, 1980b, *Lahore Urban Development and Traffic Study, Vol. 1-B—Urban Planning*, Lahore: LDA.

Lahore Development Authority (LDA) and World Bank/IDA, 1980c, *Lahore Urban Development and Traffic Study, Vol. 3—Traffic Engineering and Management Study*, Lahore: LDA.

Lahore Development Authority, 1997, *Integrated Master Plan of Lahore—Draft*, Lahore: NESPAK.

Lahore Development Authority, 2004a, *Integrated Master Plan for Lahore 2021(Draft): Volume I—Existing Scenario*, Lahore: NESPAK.

Lahore Development Authority, 2004b, *Integrated Master Plan for Lahore 2021: Volume II—Analysis and Proposals*, Lahore: NESPAK.

LaPorte, R., 1984, Administering Development in, Burki. S. and LaPorte, R. (Eds.), *Pakistan's Development Priorities: Choices for the Future*, Karachi: Oxford University Press, pp. 239–70.

LaPorte, R., 1992, Administration, in, James, W. and Subroto, R. (Eds.), *Foundations of Pakistan's Political Economy: Towards an Agenda for the 1990s*, New Delhi: Sage Publication, pp. 102–15.

Lele, S., 1991, Sustainable Development: a Critical Review, *World Development*, 19(6), pp. 607–21.

Levi, M., 1997, A Model, a Method and a Map: Rational Choice in Comparative and Historical Analysis, in, Lichbach, M. and Zuckerman, A. (Eds.), *Comparative Politics: Rationality, Culture and Structure*, Cambridge: Cambridge University Press, pp. 19–41.

Lewis, S., 1970, *Pakistan: Industrialization and Trade Policies*, Paris: Oxford University Press.

Lidskog, R., Elander, I. and Brundin, P., 2003, Towards Sustainable Urban Transportation in the European Union?, in Low, N. and Glesson, B. (Eds.), *Making Urban Transport Sustainable*, Hampshire: Palgrave Macmillan, pp. 223–39.

Liebowitz, S. and Margolis, S., 1990, The Fable of the Keys, *Journal of Law and Economics*, 33(1), pp. 1–25.

Liebowitz, S. and Margolis, S., 1995, Path Dependence, Lock-in and History, *Journal of Law, Economics and Organizations*, 11, pp. 205–26.

Lodge, M., 2003, Institutional Choice and Policy Transfer: Reforming British and German Railway Regulation, *Governance*, 16(2), pp. 159–78.

Low, N., 1997, What made it happen? Mapping the Terrain of Power in Urban Development, *Planning Theory*, 17, pp. 88–112.

Low, N., Glesson, B. and Rush, E., 2003, Making Believe: Institutional and Discursive Barriers to Sustainable Transport in Two Australian Cities, *International Planning Studies*, 8(2), pp. 93–114.

Low, N., 2003, Is Urban Transport Sustainable, in, Low, N. and Glesson, B. (Eds.), *Making Urban Transport Sustainable*, Hampshire: Palgrave Macmillan, pp. 1–22.

Low, N. and Banerjee-Guha, S., 2003, The Global Tyranny of Roads: Observations from Mumbai and Melbourne, *World Transport Policy and Practice*, 9(2), pp. 5–17.

Low, N., 2005, Gordian Note: Resisting Sustainability in Urban Transport in Australia, in, William, K. (Eds.), *Spatial Planning, Urban Form and Sustainable Transport,* UK: Ashgate Publisher, pp. 171–82.

Low, N., Glesson, B., Green, R. and Radovic, D., 2005, *The Green City, Sustainable Homes, Sustainable Suburbs*, Sydney: UNSW Press.

Lukes, S., 1977, *Essay in Social Theory*, New York: Columbia University Press.

Lukes, S., 1986, Introduction, in, Lukes, S. (Eds.), *Power*, Oxford: Basil Blackwell Ltd., pp. 1–18.

Luke, T., 1995, Sustainable Development as a Power/Knowledge System: The Problem of Governmentality, in, Fischer, F. and Black, M. (Eds.), *Greening Environmental Policy: the Politics of a Sustainable Future*, New York: Martin's Press: pp. 21–32.

Lynas, M., Six Degrees: Our Future on a Hotter Planet, Washington, D.C. National Geographic.

Macpherson, W., 1955, Investment in Indian Railways, 1845–1875, *The Economic History Review*, 8(2), pp. 177–86.

Mahoney, J., 2000, Path dependence in Historical Sociology, *Theory and Society*, 29, pp. 507–48.

Mahoney, J., 2001, Path-Dependent Explanation of Regime Change: Central America in Comparative Perspective, *Studies in Comparative International Development*, 36(1), pp. 111–41.

Malik, Z., 2006, *Transport Management in Karachi*, Karachi: Presentation in National Urban Air Quality Workshop, Karachi Mass Transit Cell, City District Government Karachi.

Malik, Z., 2004, *Mass Transit in Karachi: Moving in a Right Direction*, Islamabad: Presentation in Chartered Institute of Logistics and Transport, Pakistan Annual Seminar, Karachi Mass Transit Cell, City District Government Karachi.

March, J. and Olsen, J., 1984, The New Institutionalism: Organizational Factors in Political Life, *American Political Science Review*, 78, pp. 734–49.

March, J. and Olsen, J., 1989, *Rediscovering Institutions: The Organizational Basis of Politics*, New York: The Free Press.

Marchall, S., 1972, *Patric Geddes: Spokesman for Man and the Environment*, New Brunswick: Rutgers University Press.

Mazza, L. and Rydin, Y., 1997, Urban Sustainability: Discourses, Networks and Policy Tools, *Progress in Planning*, 47, pp. 1–74.

McKillop, A., 2004, *Oil Price Trends Through 2004–10*, available on http://www.petroleumworld.com/SunOPF112104.htm

Meadows, D., Meadows, D. and Randers, J., 1992, *Beyond the Limits: Confronting Global Collapse, Envisioning a Sustainable Future*, Post Mills: Chelsea Green Publisher.

Meakin, R., 1998, *The Lahore Public Transport Policy and Institutions Study*, Lahore: The World Bank.

Mees, P., 2000, *A Very Public Solution, Transport in the Dispersed City*, Melbourne: Melbourne University Press.

Mees, P., 2005, Privatization of Rail and Tram Services in Melbourne: What Went Wrong?, *Transport Reviews*, 25, No. 4, pp. 433–449.

Mercier, J., 2006, American Hesitations to Reduce Greenhouse Gas Emissions: an Institutional Interpretation, *International Review of Administrative Sciences*, 72(1), pp. 101–21.

Mitchell, K., 1942, India's Economic Potential, *Pacific Affairs*, 15(1), pp. 5–24.

Mitchell, R. and Rapkin, C., 1954, *Urban Traffic—A function of Land Use*, New York: Columbia University Press.

Melbourne Metropolitan Board of Works (MMBW), 1981, *Hierarchy of Roads Study: Steering Committee Final Report*, Melbourne: MMBW.

Morisugi, H., and Hayashi, Y., 2000, International Comparison of Background Concept and Methodology of Transportation Project Appraisal, *Transport Policy*, 7, pp. 73–88.

Naess, P., 2006, Cost-benefit Analysis of Transportation Investments: Neither Critical nor Realistic, *Journal of Critical Realism*, 5(1), pp. 32–60.

Nadeem, O. and Hameed, R., 2006, *Good Environmental Governance through EIA—The Case of Public Sector Development Projects in Pakistan*, Paper presented in the 12th Annual International Sustainable Development Research Conference, 6–8 April 2006, Hong Kong.

National Audit Office, 1988, *Road Planning*, House of Commons Paper 688, London: HMSO.

Newman, P. and Kenworthy, J., 1999, *Sustainability and Cities*, Washington: Island Press.

Noman, F., 2001, *Air Quality in Urban Areas in Pakistan Vs Transport Planning: Issues and Management Tools*, available on.
http://adb.org/Documents/Events/2002/RETA5937/Manila/downloads/tp_16B_qadir.pdf

Noman, O., 1997, *Economic and Social Progress in Asia: Why Pakistan did not Become a Tiger*, Karachi: Oxford University Press.

North, D., 1990, *Institutions, Institutional Change and Economic Performance*, Cambridge: Cambridge University Press.

Oil Scenarios, 2004, *Exploring the Debate over Peak Oil*, available on http://www.oilscenarios.info/index.htm

Ostrom, E., 1990, *Governing the Commons*, Cambridge: Cambridge University Press.

Ostrom, E., 1991, Rational Choice Theory and Institutional Analysis: Towards Complementarity, *American Political Science Review*, 85(1), pp. 237–43.

Owens, S., 1995, From 'Predict and Provide' to 'Predict and Prevent'? Pricing and Planning in Transport Policy, *Transport Policy*, 2(1), pp. 43–50.

Peck, J., 1996, *Technological Lock-in and the Power Source for the Motor Car*, Oxford Economic and Social History Working Papers, available on http://www.nuff.ox.ac.uk/Economics/History/

Peck, J., 1999, The Balance of Technological Transfer 1870–1914, in, Dormois, J. and Dintenfass, M. (Eds.), *British Industries Declined,* London: Routledge, pp. 114–38.

Peters, B., 1999, *Institutional Theory in Political Science: The 'New Institutionalism*, London: Biddles Ltd.

Pickrell, D., 1992, A Desire Named Streetcar: Fantasy and Fact in Rail Transit Planning, *Journal of the American Planning Association*, 58(2), pp. 158–76.

Pierson, P., 1993, When Effect Becomes Cause: Policy Feedback and Political Change, *World Politics*, 45, pp. 595–628.

Pierson, P., 2000a, Not Just What but When: Timing and Sequencing in Political Processes, *Studies in American Political Development*, 14(1), pp. 72–92.

Pierson, P., 2000b, Increasing Returns, Path Dependence and the Study of Politics, *American Political Science Review*, 94(2), pp. 251–67.

Pierson, P., 2004, *Politics in Time: History, Institutions and Social Analysis*, Princeton: Princeton University Press.

Platteau, J-P., 2004, Monitoring Elite Capture in Community-Driven Development, *Development and Change*, 35(2), pp. 223–46.

Pollack, M., 1996, The New Institutionalism and EC Governance: The Promise and Limits of Institutional Analysis, *Governance*, 9(4), pp. 429–58.

Porter, A., 2005, *Peak Oil Enters Mainstream Debate*, in BBC News, 10 June 2005.

Pucher, J., Korattyswaroopam, N., and Ittyerah, N., 2004. The Crisis of Public Transport in India: Overwhelming Needs but Limited Resources, *Journal of Public Transportation*, 7, pp. 1–20.

Qadeer, M., 1983, *Lahore: Urban Development in Third World*, Lahore: Vanguard.

Qureshi, M., 1984, *Planning and Development in Pakistan: Review and Alternatives*, Lahore: Vanguard Books Ltd.

Rhodes, R., 1981, *Control and Power in Central Local Relations*, Farnborough: Gower.

Rhodes, R., 1986, *The National World of Local Government*, London: Allen and Unwin.

Rhodes, R., 1988, *Beyond Westminster and Whitehall: The Sub-Central Government of Britain*, London: Unwin.

Richmond, J., 1990, Introducing Philosophical Theories to Urban Transportation Planning: Or Why all Planners should Practice Bursting Bubbles, *Systems Research*, 7(1), pp. 47–56.

Riker, W., 1980, Implications from the Disequilibrium of Majority Rule for the Study of Institutions, *American Political Science Review*, 75, pp. 432–47.

Robinson, G., 1991, *American Bureaucracy: Public Choice and Public Law*, Arbor: University of Michigan Press.

Rodenburg, C., Ubbeks, B. and Nijkamp, P., 2002, Policy Scenarios for Achieving Sustainable Transportation in Europe, *Transport Reviews*, 22(4), pp. 449–72.

Roque, J., 1996, The Social Dimensions of Technological Change: Reshaping Cities and Urban Life, in, Cohen, M. (Eds.), *Urban Future, Global Pressure and Local Forces*, Washington: Woodrow Wilson Center Press, pp. 171–99.

Sabatier, P., 1993, Policy Change over a Decade or More, in, Sabatier, P. and Jenkins-Smith, H. (Eds.), *Policy Change and Learning—an Advocacy Coalition Approach*, Boulder: Westview Press, pp. 13–40.

Standing Advisory Committee on Trunk Road Assessment (SACTRA), 1994, *Trunk Roads and the Generation of Traffic*, London: HMSO.

Scott, J., 2001, *Power*, UK: Polity.

Scott, R., 1994, Institutions and Organizations: Toward a Theoretical Synthesis, in, Scott. R. and Meyer, J. (Eds.), *Institutional Environments and Organizations: Structural Complexity and Individualism*, Thousand, O., CA: Sage, pp. 55–80.

Senge, P., Kleiner, A, Roberts C., Ross, R. and Smith, B., 1994, *The Fifth Discipline Field Book: Strategies and Tools for Building a Learning Organization*, New York: Doubleday.

Sewell, W., 1996, Three Temporalities: Toward an Eventful Sociology, in, McDonald, T. (Eds.), *The Historic Turn in the Human Sciences*, Ann Arbor: University of Michigan Press, pp. 245–80.

Shelvankar, K., 1940, *The Problem of India*, London: Penguin.

Shigeta, Y., 2002, *Comments on 3 Cities Environmental Investigation Results*, available on http://www.environment.gov.pk/ visited on 11 November 2006.

Small, K., 1985, Transportation and Urban Change, in, Peterson, P. (Eds.), *The New Urban Reality*, Washington: The Brookings institutions, pp. 197–224.

Smil, V., 2003, *Energy at the Crossroads, Global Perspectives and Uncertainties*, Cambridge: MIT Press.

Smith, R., 1997, Creative Destruction: Capitalist Development and China's Environment, *New Left Review*, 222, pp. 3–42.

Sohail, M., Maunder D., and Cavill S., 2006, Effective Regulation for Sustainable Public Transport in Developing Countries, *Transport Policy*, 13, pp. 177–90.

Steering Group (and Working Group of the Minister of Transport), 1963, *Traffic in Towns: A Study of the Long Term Problems of Traffic in Urban Areas, the Buchanan Report*, London: HMSO.

Strinati, D., 1995, *An Introduction to Theories of Popular Culture*, London: Routledge.

Sum, C. and Hills, P., 1998, Interpreting Sustainable Development, *Journal of Environmental Sciences*, 10(2), pp. 129–43.

Sussman, C., 1976, *Planning the Fourth Migration: The Neglected Vision of the Regional Planning Association of America*, Cambridge: MIT Press.

Sweet, s., Fligstein, N. and Sandholtz, W., 2001, The Institutionalisation of European Space, in, Sweet, s., Fligstein, N. and Sandholtz, W. (Eds.), *The Institutionalisation of European Space*, Oxford: Oxford University Press, pp. 1–28.

Szyliowicz, J., 2004, Turkey's Surface Transportation Policy and Sustainable Development, *Middle Eastern Studies*, 40(1), pp. 23–44.

The Times, 1966, Poised to Leap Forward, London, 26 February 1966.

The New York Times, 1965, Use of Fertilizer is a Main Factor, New York: 18 January 1965.

Thelen, K. and Steinmo, S., 1992, Historical Institutionalism in Comparative Politics, in, Steinmo, S., Thelen, K. and Longstreth, F. (Eds.), *Structuring Politics: Historical Institutionalism in Comparative Analysis*, New York: Cambridge University Press, pp. 1–32.

Thelen, K., 1999, Historical Institutionalism in Comparative Politics, *Annual Review of Political Science*, 2, pp. 388–92.

Thelen, K., 2003, How Institutions Evolve: Insights from Comparative Historical Analysis, in, Mohoney, J. and Diertrich R. (Eds.), *Comparative Historical Analysis in the Social Science*, Cambridge: Cambridge University Press, pp. 208–40.

Thelen, K., 2004, *How Institutions Evolve: the Political Economy of Skills in Germany, Britain, Japan and the United States*, New York: Cambridge University Press.

Thompson, P., 2004, *The Wolf at the Door, the Beginners Guide to Oil Depletion*, available on http://www.wolfatthedoor.org.uk/

Thorner, D., 1951, Capital Movement and Transportation: Great Britain and the Development of India's Railway, *The Journal of Economic History*, 11(4), pp. 389–402.

Thorner, D., 1955, The Pattern of Railway Development in India, *The Far Eastern Quarterly*, 14(2), pp. 201–16.

Tiwari, G., 2002, Urban Transport Priorities: Meeting the Challenge of Socio-Economic Diversity in Cities, a Case Study of Delhi, India, *Cities*, 19(2), pp. 95–103.

Torfing, J., 2001, Path Dependent Danish Welfare Reforms: The Contribution of the New Institutionalisms to Understanding Evolutionary Change, *Scandinavian Political Studies*, 24(4), pp. 277–309.

Traffic Engineering and Transport Planning Agency (TEPA), 1992, *Lahore Ring Road Feasibility Study*, Lahore: Mott MacDonald International Ltd.

Traffic Engineering and Transport Planning Agency (TEPA), 1993, *Budget Estimates—1994–95*, Lahore.

Traffic Engineering and Transport Planning Agency (TEPA), 1994, *Budget Estimates—1995–96*, Lahore.

Traffic Engineering and Transport Planning Agency (TEPA), 1996, *Budget Estimates—1997–98 and Revised Budget Estimates 1996–97*, Lahore.

Traffic Engineering and Transport Planning Agency (TEPA), 1997, *Budget Estimates for 1998–99 and Revised Estimates 1997–98*, Lahore.

Traffic Engineering and Transport Planning Agency (TEPA), 1998, *Budget Estimates for 1999–2000 and Revised Estimates 1998–99*, Lahore.

Traffic Engineering and Transport Planning Agency (TEPA), 2000, *Budget Estimates for 2001–2002 and Revised Estimates 2000–01*, Lahore.

Traffic Engineering and Transport Planning Agency (TEPA), 2001, *Budget Estimates for 2002–2003 and Revised Estimates 2001–02*, Lahore.

Traffic Engineering and Transport Planning Agency (TEPA), 2002, *Budget Estimates for 2003–2004 and Revised Estimates 2002–03*, Lahore.

Traffic Engineering and Transport Planning Agency (TEPA), 2003, *Budget Estimates for the Year 2004–2005 and Revised Estimates 2003–04*, Lahore.

Traffic Engineering and Transport Planning Agency (TEPA) and Japan International Cooperation Agency (JICA), 1992, *Comprehensive Study on Transportation System in Lahore: Final Report*, Lahore: TEPA.

Transport Sector Development Initiative (TSDI), 2001, *TSDI Draft Policy—Consolidate Transport Policy for all Modes, A Collective Efforts by Citizens and Government's of Pakistan*, available on http://www.pakroadusers.com/tsdi/tsdidraft.htm

United Nations (UN) Habitat, 2004, *The State of the World's Cities 2004/2005: Globalization and Urban Culture*, London: Earthscan.

United Nations Development Programme (UNDP), 2000, *Human Development Report 2000*, UNDP.

United Nations Development Programme (UNDP), 2006, *Human Development Report 2006*, available on http://hdr.undp.org/hdr2006/pdfs/report/HDR06-complete.pdf visited on 28 November 2006.

United Nations Framework Conventions on Climate Change (UNFCCC), 2001, *Greenhouse Gas Inventory Database', United Nations Framework Convention on Climate Change*, available on http://ghg.unfccc.int/

Vakil, C., 1944, Railways and Roads in India, *Annals of the American Academy of Political and Social Science*, 233, pp. 187–92.

Vasconcellos, E., 2001, *Urban Transport, Environment and Equity: The Case for Developing Countries*, London: Earthscan.

Vigar, G., 2002, *The Politics of Mobility: Transport, the Environment and Public Policy*, London: Spon Press.

Volvo, 1980, *Model Urban Transport System Lahore: Summary of Findings and Recommendations*, Lahore.

Wallace, H., 1980, *Budgetary Politics: The Finances of the European Communities*, London: George Allen and Unwin.

Wardle, R. and Smith, I., 2004, Modelled Response of the Australian Monsoon to Changes in Land Surface Temperatures, *Geophysical Research Letters*, 31, LI6205.

Waterston, A., 1963, *Planning in Pakistan: Organization and Implementation*, Baltimore: The Johns Hopkins Press.

Weir, M. and Skocpol, T., 1985, The State Structures and the Possibilities for 'Keynesian' Responses to the Great Depression in Sweden, Britain and the United States, in, Evans, P. (Eds.), *Bringing the State Back*, Cambridge: Cambridge University Press, pp.107–63.

Wesolowsky, T., 2005, *When Oil Peaks*, Asia Times, 26 January 2005.

Whitelegg, J., 1993, *Transport for a Sustainable Future: The Case of Europe*, London: Belhaven Press.

Whitelegg, J., 1997, *Critical Mass: Transport, Environment and Society in the Twenty-first Century*, London: Pluto Press.

Wilson, G., 1966, Introduction, in, Wilson, G., Bergmann, B., Hirsch, L. and Klein, M. (Eds.), *The Impact of Highway Investment on Development*, Washington DC: The Brookings Institutions, pp. 1–16.

Wilson, D., Game, C. Leach, S., and Stoker, G., 1994, *Local Government in the UK*, Basingstoke: Macmillan.

Wilson, R., 2001, Assessing Communicative Rationality as a Transportation Planning Paradigm, *Transportation*, 28(1), pp. 1–31.

Woodlief, A., 1998, The Path-Dependent City, *Urban Affairs Review*, 33(3), pp. 405–37.

World Bank, 1994, *World Development Report 1994: Infrastructure for Development*, New York: Oxford University Press.

World Bank, 1996, *Sustainable Transport: Priorities for Policy Reform*, Washington D.C.

World Bank, 2003, *Traffic Fatalities and Economic Growth*, Policy Research Working Paper 3035, Washington D.C.

World Bank, 2005, *World Development Indicators*, available on: http://web.worldbank.org/WBSITE/EXTERNAL/DATASTATISTICS/0, contentMDK:20523710~hlPK:1365919~menuPK:64133159~pagePK:64133150~piPK:64133175~theSitePK:239419,00.html visited on 28 November 2006.

World Bank, 2006a, *Options for Mass Rapid Transit (MRT) under an Integrated Urban Transport Framework*, Karachi: Presentation in National Workshop on Urban Air Quality.

World Bank, 2006b, Pakistan—*Punjab Large Cities DPL Project—Programme Information Document (PID), Concept Stage*, Washington D.C.

World Bank, 2006c, *Pakistan Strategic Country Environmental Assessment; Volume II, Technical Annex*, Islamabad: South Asia Environment and Social Development Unit.

World Commission on Environment and Development (WCED), 1987, *Our Common Future*, Oxford: Oxford University Press.

World Health Organisation (WHO), 2004, *World Report on Road Traffic Injury Prevention*, WHO.

Yin, R., 1994, *Case Study Research: Design and Methods*, New York: Sage.

Zahidi, T., 1998, *The Politics of Sustainable Urban Environmental Management in Islamabad*, Unpublished PhD Thesis: University of East Anglia, UK.

WEBSITES CITED

http://dawn.com/ (*Dawn*, 28 March 2002, Internet edition)

http://www.punjab.gov.pk/p-d/h_roads.htm visited on 10 December 2005

http://bbc.co.uk/urdu visited on 16 December 2005

http://en.wikipedia.org/wiki/Public_policy visited on 10 April 2006
http://en.wikipedia.org/wiki/Rate_of_return_on_investment visited on 15 May 2006
http://globalis.gvu.unu.edu/country.cfm?country=PK&indicatorid=0 visited on 5 April 2005
http://irfca.org/faq/faq-hist.html#first visited on 13 October 2005
http://news.bbc.co.uk/1/hi/business/4077802.stm visited on 15 June 2005
http://siteresources.worldbank.org/PAKISTANEXTN/Resources/293051-1114424648263/Session-VII-Farrukh.pdf visited on 30 April 2005
http://www.adb.org/Documents/TARs/PRC/tar-prc-38174.pdf#search='TAR%20PRC%2038174%20TECHNICAL%20ASSISTANCE%20TO%20THE%20PRC' visited on 10 April 2006
http://www.atimes.com/atimes/Global_Economy/GA26Dj04.html visited on 13 June 2005
http://www.enercon.gov.pk/ferts/ visited on 22 June 2005
http://www.enercon.gov.pk/issues.asp visited on 5 June 2005
http://www.factbook.net/EGRF_Regional_analyses_AsiaPacific.htm visited on 20 January 2003
http://www.hdip.com.pk visited on 25 June 2005
http://www.iepsac.org/papers/p04a11.htm visited on 27 October 2005
http://www.indianchild.com/british_empire_in_india.htm visited on 5 July 2005
http://www.ipcc.ch visited on 27 May 2005
http://www.met-office.gov.uk/research/hadeleycentre visited on 15 June 2005
http://www.met-office.gov.uk/research/hadeleycentre/models/carbon_cycle/index.html visited on 15 June 2005
http://www.mpnr.gov.pk/fosilfuel.php visited on 22 June 2005
http://www.nespak.com.pk/news/newsdetail.asp?id=365 visited on 6 January 2006
http://www.nespak.com.pk/news/newsdetail.asp?id=497 visited on 6 January 2006
http://www.nuff.ox.ac.uk/Economics/History/ visited on 15 April 2005
http://www.nust.edu.pk/R&D.htm visited on 30 June 2005
http://www.oilscenarios.info/index.htm visited on 15 June 2005
http://www.pakistaneconomist.com/database1/macro-economy/macro-economy-77.htm visited on 10 March 2005
http://www.peakoil.net/ visited on 15 June 2005
http://www.petroleumworld.com/SunOPF112104.htm visited on 15 June 2005
http://www.pmupunjab.gov.pk/index.php?option=com_content&task=view&id=17&Itemid=47 visited on 10 January 2006
http://www.pmupunjab.gov.pk/index.php?option=com_content&task=view&id=17&Itemid=4gttp://www.pmupunjab.gov.pk/index.php?option=com_content&task=view&id=17&Itemid=47 visited on 12 December 2005
http://www.statpak.gov.pk/depts/pco/statistics/pop_education/pop_education_sex.html visited on 10 January 2005
http://www.statpak.gov.pk/depts/fbs/publications/Year%20Book%202005/social%20and%20Culture/19.5.pdf visited on 5 November 2006
http://www.statpak.gov.pk/depts/fbs/publications/Year%20Book%202005/population/16.10.pdf visited on 5 November 2006
http://www.uettaxila.edu.pk/civilPGprog.htm visited on 30 June 2005
http://www.un.org/esa/sustdev/documents/agenda21/english/agenda21toc.htm visited on 15 August 2003
http://www.victorianweb.org/history/empire/timeline.html visited on 5 July 2005
http://www.wolfatthedoor.org.uk/ visited on 15 June 2005

http://www.worldbank.org.pk/WBSITE/EXTERNAL/COUNTRIES/SOUTHASIAEXT/PAKISTANEXTN/0, contentMDK:20131431~menuPK:293057~pagePK:1497618~piPK:217854~theSitePK:293052,00.html visited on 10 May 2005

http://www.jang.com.pk/ (Daily *Jang*, 10 June 2003, Internet edition, Urdu Newspaper)

http://www.nawaiwaqt.com.pk/ (*Nawa-i-waqt*, 14 May 2003, Internet edition, Urdu Newspaper)

http://nation.com.pk (*The Nation*, 23 January 2003, Internet edition)

http://www.thenews.com.pk/ (*The News*, 16 October 2002, Internet edition)

http://www.adb.org visited on 15 August 2005

http://www.enercon.gov.pk visited on 25 January 2003

http://www.oed.com online Oxford English Dictionary

http://www.pakroadusers.com/tsdi/index.htm visited on 15 January 2005

http://www.punjab.gov.pk visited on 18 December 2005

http://www.sdpi.org visited on 7 July 2005

http://www.transparency.org visited on 15 May 2006

http://www.environment.gov.pk/ visited on 11 November 2006.

Index

C

D

E

J

K

L

O

P

Q

R

U